SpringerBriefs in Computer Science

T0213904

For further volumes:
http://www.springer.com/series/10028

Sanaa Taha · Xuemin Shen

Secure IP Mobility
Management for VANET

Sanaa Taha
Information Technology Department
Faculty of Computers and Information
Cairo University
Orman, Giza
Egypt

Xuemin Shen
Department of Electrical and Computer
  Engineering
University of Waterloo
Waterloo, ON
Canada

ISSN 2191-5768   ISSN 2191-5776   (electronic)
ISBN 978-3-319-01350-3   ISBN 978-3-319-01351-0   (eBook)
DOI 10.1007/978-3-319-01351-0
Springer Cham Heidelberg New York Dordrecht London

Library of Congress Control Number: 2013944767

Printed on acid-free paper

Springer is part of Springer Science+Business Media (www.springer.com)

Preface

With the proliferation of IP-based vehicular applications, such as Internet access and Infotainment, the underlying mobility management protocols for Vehicular Ad Hoc Networks (VANETs) have been recently witnessed noticeable improvements. Mobility management protocols for VANETs are envisioned to support mobile nodes (MNs), i.e., vehicles, with seamless communications, in which service continuity is guaranteed while vehicles are roaming through different RoadSide Units (RSUs).

In this brief, we present the challenges and solutions for VANETs' security and privacy problems occurred in three mobility management protocols, namely Mobile IPv6 (MIPv6), Proxy MIPv6 (PMIPv6), and Network Mobility (NEMO). In Chap. 1, we first give an overview of the concept of the vehicular IP-address configurations as the prerequisite step to achieve mobility management for VANETs, and then review the current security and privacy schemes applied in the three mobility managements. In Chap. 2, we propose an anonymous and location privacy-preserving scheme (ALPP) for the MIPv6 protocol. Based on the onion routing and anonymizer, ALPP involves two complementary subschemes: anonymous home binding update (AHBU) and anonymous return routability (ARR), and it is more suitable to be applied for high-mobile networks, such as the VANETs. In Chap. 3, we discuss the multihop authentication problem in PMIPv6-based VANET and propose EM^3A, a novel mutual authentication scheme that guarantees the authenticity of both MN and the relay node (RN). EM^3A scheme relies on the symmetric polynomial authentication, and it thwarts authentication attacks, including Denial of service (DoS), collusion, impersonation, replay, and man-in-the-middle attacks. For a PMIP domain with (n) points of attachment and a symmetric polynomial of degree (t), our scheme achieves ($t \times 2^n$)-secrecy, whereas the existing symmetric polynomial-based authentication schemes achieve only t-secrecy. In Chap. 4, we deal with the physical-layer location privacy attacks in the NEMO-based VANETs scenario and propose a new physical-layer location privacy scheme, the fake-point cluster-based scheme, to prevent attackers from localizing users inside NEMO-based VANET hotspots. The proposed scheme can: (1) confuse the attackers by increasing the estimation errors of their Received

Signal Strength (RSSs) measurements, and (2) prevent attackers' monitoring devices from detecting the user's transmitted signals. Finally, we conclude the Brief and provide future research directions in Chap. 5.

April 2013   Sanaa Taha
                                                                 Xuemin Shen

Contents

Chapter 1
Introduction

Nowadays, academic and industry are concerning about supporting vehicular networks with seamless communications, by which mobile nodes (MNs), i.e., vehicles, are allowed to roam through different RoadSide Units (RSUs) while keeping their active IP sessions [1]. Achieving such communications is required to support IP-based applications, such as video streaming and infotainment. Mobility management protocols, such as Mobile IPv6 (MIPv6) and the NEtwork MObility (NEMO) protocols, are envisioned to implement seamless communications in mobile networks, including Vehicular Ad Hoc Networks (VANETs).

The integration of the current mobility management standards, namely Mobile IP (MIP), and VANET is performed through three functionalities: (1) vehicular IP address configuration, to configure a unique and permanent IP address to a vehicle, (2) IP mobility mechanisms, to manage the mobility of the configured vehicle while roaming across many networks, and to keep its connectivity to the Internet, and (3) forwarding IP packets among vehicular networks, to forward the vehicle's IP packets through vehicle routing schemes. In this brief, we focus on securing the second functionality, specifically, achieving security and privacy for IP mobility employed with VANET. In this chapter, In addition to the IP mobility, we briefly review the vehicular IP auto-configuration address, science it is a prerequisite not only for IP mobility, but also for address-based routing protocols in VANETs.

1.1 Vehicular IP Address Configurations

In VANETs, the challenges of multihop nature and the lack of multicast communications prevent vehicles employing traditional IP stateless and stateful auto-configuration schemes to assign unique IP addresses. Moreover, auto-configuration schemes for Mobile Ad hoc Networks (MANETs) do not work for VANETs [2] due to the high delays. Therefore, the configuration of IP address in VANETs is a

S. Taha and X. Shen, *Secure IP Mobility Management for VANET*,
SpringerBriefs in Computer Science, DOI: 10.1007/978-3-319-01351-0_1,
© The Author(s) 2013

key research issue, and current IP address configuration are classified into centralized, distributed, and geographical based schemes. Both centralized and distributed
schemes base on the Dynamic Host Configuration Protocol (DHCP), while the geographical based scheme relies on the vehicles' geographic locations.

1.1.1 Distributed IPv6 Vehicular Configuration

The Vehicular Address Configuration (VAC) [3] is the first to depend on the VANET
topology and the DHCP enhancement. Instead of having a DHCP server for the whole
network, VAC creates a leader chain among the communicating vehicles. Each leader
works as a DHCP server to support vehicles in its coverage area with an IP address.
Employing a SCOPE, that is a number of hops, each leader determines the number
of vehicles under its coverage area. An ordinary vehicle may become a leader if the
number of hops to its leader is larger than a max-threshold of the leader's SCOPE,
and a leader may change to an ordinary vehicle if many leaders are located near
to each other. Within a defined coverage area, VAC provides unique IP addresses,
however, a Duplicate Address Detection (DAD) scheme is required to detect any
duplicated IP addresses outside this area.

In addition, the Cluster-based Addressing Scheme in VANET (CAVET) [4]
divides the VANET's area into clusters to guarantee the scalability of the networks
with high-mobile vehicles. CAVET is proposed mainly for the ad hoc structure in
VANET (V2V communications), in which cluster heads are responsible for propagat-
ing vehicles' packets. In addition, the Regional-based Auto-Configuration Protocol
Association with Coding Architecture for VANETs (RAPACA) [5] is used for V2I
communications, therefore, it is used along with the CAVET protocol to support
V2V2I communications in VANETs. RAPACA divides the region into clusters, and
assigns a unique code for each area. This scheme also designs a new IPv6 distribution scheme, in which the 16 bits of the host IP present the vehicle's home network
and the last 16 bits of the network prefix present the cluster ID (CID), as depicted
in Fig. 1.1. The DAD scheme is no longer needed because the vehicle' s home network part guarantees the uniqueness of the IP address, and hence the signaling delay
decreases. However, a scalability problem occurs because the length of the host identity decreases to 48 bits, and hence a maximum of 2^{48} vehicles can be assigned with
IP addresses inside each cluster.

0-63 Network Prefix		64-127 Host ID	
Site Network	CID	Home ID	Host ID
48 bits	16 bits	16 bits	48 bits

Fig. 1.1 IPv6 distribution structure [5]

1.1.2 Centralized IPv6 Vehicular Configurations

The centralized Address Configuration (CAC) protocol [6] uses a centralized DHCP to support vehicles in urban areas with unique global IP addresses. The address request transmitted from the vehicle to the DHCP server is relayed by the RSUs distributed along the road. Centralized schemes depend on a single point of failure, the DHCP server, therefore, they are more complex than distributed schemes. However, the signaling delay is much lower than that in the distributed.

1.1.3 Geographical-Based IPv6 Vehicular Configurations

Unlike both centralized and distributed schemes, which are stateful configurations, the geographical based schemes are stateless auto-configurations. The Geographically Scoped Stateless Address Configuration (GeoSAC) [7] adapts the existing IPv6 Stateless Address Auto-configuration (SLAAC) scheme by extending the IPv6 link to a specific geographical area controlled by a point of attachment, i.e., Access Point (AP).

Each point of attachment, periodically broadcasts IPv6 Router Advertisement (RA) messages that contain an IPv6 prefix to be used by mobile nodes, i.e., vehicles, located within a well-defined geographical area. When a node receives an RA message, the node first applies geographic filtering and then forwards the message to other vehicles in its geographical area. Consequently, a multihop path is created among the vehicles, which create their IPv6 addresses by appending their network identifier, derived from their MAC address, to the received IPv6 prefix. Finally the vehicle applies a DAD scheme to ensure the uniqueness of its derived IPv6 address.

1.2 IP Mobility Management for Vehicular Networks

The first trend to manage vehicles' mobility was through the MIP protocols, due to their standardizations and widely deployment. Involving location and handover managements, the IP mobility management for VANETs tracks the vehicle's location to enable packet reception. In addition, IP mobility supports seamless handover, also called seamless communications, by which the vehicle's connection to the infrastructure is kept active while the vehicle roams to different points of attachment. When using this time-restricted handover process, both mobile node and service provider have some benefits, including low cost, wide coverage, and high bandwidth.

Having the same goal of supporting global Internet connectivity and IP-based applications, mobility management protocols [8] are classified into host-based and network-based mobility. In host-based mobility management protocols, such as Mobile IPv6 (MIPv6) [9], the mobile node manages its own mobility, whereas in network-based mobility protocols, such as Proxy MIPv6 (PMIPv6) [10], the mobility of an MN is managed by network entities, such as ARs, without involving the MN. In addition, the network mobility protocol, also called NEMO [11], is an extension of the MIPv6 protocol to manage the mobility of the moving network as one unit.

1.2.1 Mobile IPv6

With Mobile IPv6 (MIPv6), depicted in Fig. 1.2, each Mobile Node (MN) has two different IP addresses: a Home Address (HoA) and a Care of Address (CoA). The HoA is the original MN's address that is configured by the MN's Home Agent (HA), which is a router located in the MN's home network. The CoA is acquired from a Foreign Gateway (FG), which is a router located in the visited network.

For an MN, to implement seamless communications when it roams to a foreign network, it employs the mobile IPv6 control messages: home binding update, transmitted to the MN's Home Agent (HA), and return routability, transmitted to the MN's Correspondent Nodes (CNs). By sending these control messages, an MN informs both its HA and its CNs about its current location, which is represented by its CoA. Therefore, the roaming MN can receive any subsequent messages, destined for its HoA, at this CoA. Both HA and CN create bindings between the MN's home address and CoA, and then transmit any subsequent messages to this CoA instead of to the MN's HoA.

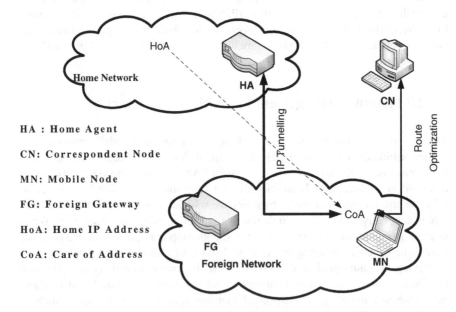

HA : Home Agent

CN: Correspondent Node

MN: Mobile Node

FG: Foreign Gateway

HoA: Home IP Address

CoA: Care of Address

Fig. 1.2 Roaming among mobile IPv6 heterogeneous networks

However, MIPv6 protocol increases the handover latency by frequently sending the control messages to an MN's HA. Therefore, other MIP variants, such as the Hierarchial MIPv6 (HMIPv6) [12] and Fast MIPv6 (FMIPv6) [13], are proposes to decrease delays and increase network performance. HMIPv6 decreases the round trip time delay in MIPv6 by choosing a Mobile Anchor Point (MAP), close to the MN, to work as the MN's HA. In addition to its CoA that is known as Regional CoA (RCoA), the MN acquires another address called On-link CoA (LCoA) from the MAP that manages the MN's area. Furthermore, this LCoA address changes as the MN moves to a new MAP's domain. Instead of sending frequent BU messages to its far-off HA, the MN sends local BUs to the MAP in order to bind its LCoA with its RCoA. In addition, the MN sends a BU message to its HA to bind its HoA and RCoA. The BU messages to MAP are transmitted frequently, whereas there is no need to send those messages to the MN's HAAs as long as the MN is located in the same MAP domain.

Furthermore, FMIPv6 creates a tunnel between an MN's previous AR (PAR) and new AR (NAR). This tunnel is used to transmit MNs' messages from PAR to NAR during the handover time, and it is disconnected after the MN fully moves to the new subnet. When the MN detects its link-layer movement to a new subnet, it sends a router solicitation proxy message to its PAR asking for the identification of the new subnet, and then sends a fast BU message to its PAR, after creating a new CoA (NCoA) based on the new subnet identification. The PAR sends a handover initiate (HI) message to the NAR in order to create a tunnel between PAR and NAR. To reply to the HI message, the NAR first carries a DAD mechanism for the NCoA to guarantee its uniqueness.

1.2.2 Proxy Mobile IPv6

As a network-based mobility management protocol, Proxy Mobile IPv6 (PMIPv6) [14] employs network's access points to manage the MN's mobility. In addition, as a localized protocol, with PMIPv6, the MN moves within its local mobility domain (LMD) without changing its IP address. The LMD contains one or more Local Mobility Anchors (LMAs) that work as an MNs' home agents, and a group of Mobile Access Gateways (MAGs) that send the mobility signallings to LMAs on behalf of MNs. When the MN joins the LMD, it sends a Router Solicitation (RS) to its directly attached MAG, which sends a proxy binding update (PBU) message to the LMA. After authorizing the MN, a tunnel between the the LMA and the MAG is crated, and a Proxy Binding Acknowledgement (PBA) message containing the MN's Network Prefix(es) (NPs) is sent to the MAG. Working as a proxy, the MAG forwards the PBA to the MN by sending a router advertisement message. Employing the NPs, the MN creates its IP address, and then uses a DAD scheme to guarantee the uniqueness of its configured IP address. Therefore, the MN moves among different MAGs inside the LMD using its configured address and without detecting its movements, which indeed are detected by the attached MAGs and the LMA. Figure 1.3 shows the PMIPv6 operations.

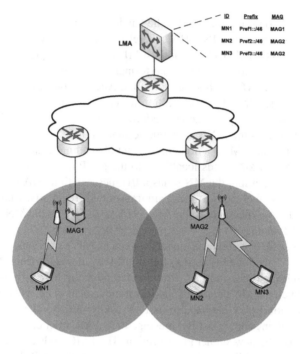

Fig. 1.3 Proxy Mobile IPv6 [14]

1.2.3 Network Mobility

The NEMO basic support (NEMO BS) protocol [15] is an extension of the mobile IP protocol [16, 17], in which a Mobile Router (MR) works as an MN to manage, not only its own mobility, but also an entire network of Mobile Network Nodes (MNNs) located under its coverage. Therefore, in NEMO-BS, mobility management functionalities are performed by the MR rather than the MNNs, which only implement the basic IP protocol without being aware of the entire network mobility.

As the first step to create its network, the MR takes the responsibility for managing the mobility of the entire network by periodically broadcasting its Mobile Network Prefixes (MNPs), acquired from the MR's home network. To join the network, each MNN selects a distinct MNP to be its address in the moving network. When moving out from its home network, the MR acquires a new Care of Address (CoA) from the Foreign Agent (FA), located in the foreign network, and sends home binding update messages to its Home Agent (HA) to bind its Home Address (HoA) with its new CoA. The Transmitted binding update messages can be in explicit or implicit mode. The explicit mode appends the MR's MNPs to the transmitted binding update messages. The implicit mode does not append the MNps because the MR and its HA implement a dynamic routing protocol, which facilitates the HA's ability to identify the MR's MNPs. Consequently, the whole network's movement is managed

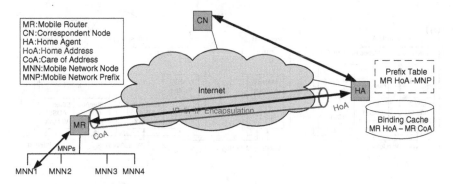

Fig. 1.4 NEMO-based VANET

by this MR. To store the bindings between the MR's HoA with CoA and MNPs, the HA stores a prefix table. Finally, a tunnel between MR's CoA and HA is created; therefore, messages transmitted between MNNs and Correspondent Nodes (CNs) are sent first to the HA, as illustrated in Fig. 1.4.

In supporting the MNNs with the required mobility, the NEMO BS reduces the signaling overhead and the mobility costs, compared to the MIP protocol. However, NEMO BS is designed to support a single-hop mobile network, in which a direct communication between an MR and the Internet access router is formed. Therefore, to support the multihop Vehicle-to-Vehicle-to-Infrastructure (V2V2I) communications, the integration of NEMO with VANET, namely NEMO-based VANET, has two roles: (1) applying session continuity and global Internet access via NEMO BS, and (2) supporting multihop communication via V2V2I routing schemes such as georouting [18, 19].

To implement the integrate between the NEMO BS and VANET, two approaches have been defined [20, 21]: MANET-centric and NEMO-centric. In the MANET-centric approach, V2V2I multihop communications are supported by implementing a NEMO BS protocol that is run on top of a MANET routing protocol. The advantage of this approach is the separation of the MANET routing from the NEMO BS, as depicted in the protocol stack shown in Fig. 1.5a. On the other hand, the NEMO-centric approach supports multihop communications by implementing at least one NEMO mobile routing scheme in the vehicles that form the V2V2I path between the MR in the mobile network and the infrastructure. In addition to act as an MR, each OBU, in the intermediate V2V2I communication path, works also as a relay for the MNNs. MANET routing protocols can be used in the NEMO-centric approach to optimize the routing paths resulting from the NEMO BS protocol. Figure 1.5b shows the protocol stack for the NEMO-centric approach, which is more appropriate for nested NEMO and hierarchical structured networks, whereas the MANET-centric approach is more suitable for our scenario, wherein the ad-hoc structure is implemented in multihop communication. Table 1.1 compares the two approaches.

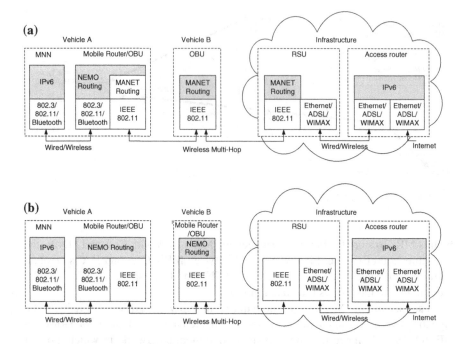

Fig. 1.5 NEMO-based VANET integration approaches. **a** MANET-Centric approach. **b** NEMO-Centric approach

Table 1.1 MANET-Centric and NEMO-Centric comparison

MANET-centric	NEMO-centric
Less mobility cost	High cost
Simple V2V2I implementation	Complex implementation
Self MNP delegation of OBU/MR	Hierarchical MNP delegation
Less signaling overhead	High signaling
Ad-hoc domain	Infrastructure domain
Georouting protocols can be	Hierarchical topology can
used for multihop communications	not use georouting protocols

1.3 Securing IP Mobility for VANETs

Integrating mobility management with VANET as a kind of mobile and heterogeneous networks [22], poses challenges—a roaming vehicle changes its point of
attachments frequently, and this change causes the network topology to be adapted
abruptly. In addition, the security and privacy preservations are critical challenges
in this mobility management-VANETs integration. Passengers prefer using the IP-
based applications when spending at least 2 h per day in vehicles, however, most of
them concern about their privacy. Relying on the IPSec protocol, Mobile IP (MIP)

standards support limited authentication and data privacy. For example, the industry reports the lack of wide adoption of the MIPv6 protocol due to the problems with interoperability in IPSec implementations. In addition, implementing IPSec with VANETs increases the vehicle's handover delay, which in turn prevents seamless communications for real-time applications. In this section, the existing security schemes applied for MIP protocols, namely MIPv6, PMIPv6, and NEMO BS, are reviewed.

1.3.1 Securing MIPv6

To secure the MN's communications to the HA and CNs, the MIPV6 protocol defines both the IPSec protocol and the return routability procedure. The IPSec protocol is an Internet protocol, which was implemented originally to support wired network with confidentiality and authentication security services. With the introduction of mobility, the Internet Engineering Task Force (IETF) has updated the IPSec to support the Internet mobility protocols [23]. The IPSec protocol consists of two protocols: Encapsulating Security Payload (ESP) protocol, and Authentication Header (AH) protocol. Moreover, the IPSec is implemented in two different modes: the transport mode and the tunnel mode. The IPSec protocol is used by the mobile IPv6 protocol to secure the signalling data transmitted between the mobile node and the home agent. Due to its static policy configurations, the IPSec protocol alone cannot support the mobile IPv6 protocol with the required security mechanisms. However, the mobile IPv6 protocol requires changing the policy as a consequence of the mobility feature. Therefore, the return routability procedure is introduced to achieve both security and privacy of the mobile node.

Two goals of the return routability procedure are defined: (1) authenticating the mobile node to the correspondent node, and (2) constructing a shared key between them. As depicted in Fig. 1.6, this procedure consists of four transmitted messages transferred between the two parties, namely an MN and CN.

1. Home Test Init (HoTI) Message: This message is sent by the roaming mobile node to the correspondent node through the home agent using the IP tunneling (1a in the figure). It contains a random number generated by the mobile node called home init cookie. It is protected using the IPSec ESP protocol in the tunnel mode. The inner source address of this message is the mobile node's home address, the inner destination address is the correspondent-node's address, the outer source address is the mobile node's care of address, and the outer destination address is the home agent's address.
2. Care of-Test Init (CoTI) Message: This message is sent by the mobile node directly to the correspondent node and it contains another random number called care-of Init Cookie (1b in the figure). It does not contain any security mechanism to protect the transmitted data.
3. Home Test Message: This message is sent by the correspondent node to the mobile node, through the mobile node's home agent, using IP tunneling (2a in the figure). The inner source address of this message is the correspondent node's address, the

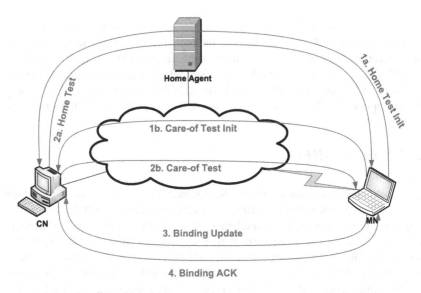

Fig. 1.6 Return Routability Procedure

inner destination address is the mobile node's home address, the outer source address is the mobile node's home agent, and the outer destination address is the mobile node's CoA. It contains three values used to construct a secret key between the mobile node and the correspondent node. The first value, the home init cookie, is the same random value, which the mobile node sends in the home test init message. The second value, the home keygen token, is a secret value generated using the HMAC-SHA1 algorithm as follows:

$$Homekeygentoken = First(64, HMAC - SHA1(K_{cn}, (HoA|nonce|0)))$$
$$(1.1)$$

Where K_{cn} is a 20-octet secret key and nonce is 64-bit secret value. Both K_{cn} and the nonce are known only by the correspondent node. The home keygen token is the first 64 bits of the HMAC-SHA1 function, which takes the concatenation of the mobile node's home address (HoA), the secret nonce, and a zero-octet as inputs and uses the K_{cn} as the secret key of the function. The third value, the home nonce index, is sent instead of sending the nonce itself as it is a secret value and cannot be sent to the mobile node.

4. Care of test (CoT) message: This message is sent by the correspondent node directly to the mobile node without going through the home agent (2a in the figure). The source address of this message is the correspondent node's address and the destination address is the mobile node's CoA. It contains three different values: care of init cookie, care of keygen token, and care of nonce index. The care of init cookie is the same value which the mobile node sends to the correspondent node in the care of test init message. The care of keygen token is another secret value generated as follows:

$$Careofkeygentoken = First(64, HMAC - SHA1(K_{CN}, (CoA|nonce|1)))$$
$$(1.2)$$

The care of keygen token value is generated similarly to the home keygen token generation, except that the care of address is used instead of the home address and one-octet is appended instead of zero-octet.

Using both the home keygen and the care of keygen tokens, the MN and the CN construct a temporary shared key, called a binding management key, K_{bm}. This shared key is used to secure the binding update messages transmitted between the MN and the correspondent node. The binding management key is constructed as follows:

$$K_{bm} = SHA1(homekeygentoken|care - ofkeygentoken).   (1.3)$$

1.3.2 Securing PMIPv6

The PMIPv6 protocol employs the IPSec to secure the communication between the LMA and MAGs, however, there is no standard security scheme defined for the communications between an MN and MAGs. In order to authenticate an MN that roams to the PMIPv6 domain, the Internet Authentication, Authorization, and Accounting (AAA) servers are used with the help of the LMA and MAGs that forward the MN's credentials to the AAA servers. However, because of the large delay, this method of communication, between MN and AAA, is not suitable for such a localized mobility protocol, PMIPv6, whose goal is to decrease the transmission delay in seamless communications. In [24], a localized authentication and billing scheme based on the hash-chain value is proposed to authenticate the roaming MN in PMIPv6 domain. A key agreement scheme, between the MN and the AAA server, creates the seed of the hash-chain, and delegates the authentication to the LMA in order to create the hash-chain on behalf of the MN. Using a hash-chain decreases the communication distances between the MAGs and AAA, as well as creates an electronic currency to be used by the MNs.

1.3.3 Securing NEMO BS

Among the mobility management protocols, the NEMO BS protocol is more suitable for VANET's IP-based applications. Therefore, many security considerations have been proposed to protect vehicles when the NEMO protocol is employed. Using the AAA servers deployed in each subnet along the Internet, and the Foreign AAA (FAAA) server, the scheme in [25] employs the PANA protocol [26] to authenticate the MR to its Home AAA (HAAA) when roaming to a visited network. This authentication scheme is implemented by the MR when sending authentication request to its HA, hence, it requires $300\,s$, the average delay for the transmitted request. In [27], the MR can be authenticated by the FAAA server rather than the HAAA server, in order to decrease the authentication time. A cost function is calculated to decide whether to use FAAA or the HAAA. In addition, the LMAM scheme [28] is an authentication scheme proposed for the MR in VANETs to locally authenticate itself to the FAAA. In this scheme, the FAAA server authenticates the MR by this MR's MAC address and without referring to the MR's HAAA. Due to the requirement of storing all MRs' MAC addresses, this scheme can only be used in small networks, and it does not work for Internet-based VANET's applications.

Although it is not standardized yet, the NEMO route optimization security has been greatly studied. In [29], a route optimization scheme based on certificated nodes has been proposed to be employed in NEMO-based networks. In the proposed scheme, using the public-key certificates, the MRs and correspondent routers (CRs) mutually authenticate each other as well as authenticate the MNPs. Assuming a pre-assigned public key infrastructure (PKI), the proposed scheme uses a trusted third party to construct the PKs and the certificates for all uses.

Furthermore, [30] mitigates the high latency problem that is found in traditional route optimization schemes, by proposing a new secure NEMO route optimization, called SeNERO. With the assumption of the ease of construction of a PKI in aeronautical systems, SeNERO creates a mutual authentication as well as a direct path between an MR and a CR. Every router, including the MR and the CR, uses a hierarchical certificate authority system to prove its authenticity, by appending its outgoing messages with a certificate chain, which includes the router's certificates starting from it and ending at the trusted root Certificate Authority (CA). SeNERO achieves two levels of authentications, initial and subsequent authentications. In the initial authentication, both the MR and CR share a secret key, K_s, which is constructed by CR on the MR's demand. Afterwards, both MR and CR exchange their certificate path protected by the K_c and their signatures to prove their authenticity and their ownership for the MNPs. In the subsequent authentication, which is performed after an MR's handover occurs, the CR creates a new K_s for the roaming MR and uses the key to protect the transmitted BU and BA messages. Unlike the initial authentication, both certificate and signature operations are not performed in subsequent authentication.

Although SeNERO seems to achieve a high level of security by using public key operations, it suffers from a security problem: transmitting the shared key in clear form without encrypting it. The attacker can easily eavesdrop the transmitted key and break the system. Moreover, the scheme is based on a certificate chain that relates the MR to the root CA without considering the mobility feature of the MR. When the MR moves to a new network, the certificate chain needs to be changed to adapt to the MR's new location. In addition, due to using digital signature, SeNERO faces a problem of large handover delays, in which the delays reach $100\,s$ for each authentication level.

1.4 Mobility Management and Security Challenges

Due to the unique characteristics of VANETs that conflict with the IP mobility, securing mobility management protocols for VANETs is a challenge, in which traditional security algorithms cannot be used. As a kind of mobile network, the mobility management-based VANET's security requirements are described herein:

1. **Authentication:** Each mobile node, i.e., vehicle, must be able to authenticate itself to both its home agent and the correspondent node communicating with it. Moreover, the mobile node requires the ability to authenticate the source of its received data. The mutual authentication requirement protects mobile IPv6 users from impersonation and masking attacks.
2. **Communication privacy:** Communication among the mobile node, the home agent, and the correspondent node should be confidential, and an intermediate eavesdropper should not learn the content of the transmitted data.
3. **Message integrity:** Active attacks should be prevented by message integrity mechanisms, by which the mobile node guarantees that the received data is not modified during transmission.
4. **Anonymity and location privacy:** The mobile node must guarantee its anonymity in the network. The anonymity of a network is the ability to hide a specific item among a group of similar items, and location privacy is the ability to prevent tracking a user's mobility. Anonymity and location privacy prevent a traffic analysis attacks from learning the identity or the location of the mobile node.
5. **Payment protocol:** The roaming mobile nodes always use the foreign networks' resources, therefore a payment protocol is required to charge the mobile node. The payment protocol must be secure in protecting the location privacy of the mobile node; moreover, the mobile node must authenticate the network operator before paying for the service.

1.4.1 Mobility Management Challenges

Specific to the integration of IP mobility with VANETs, the following challenges occur [31]:

1.4.1.1 Seamless Mobility with High Speed VANETs

Due to the high speeds of vehicles and hence the increase of number of handovers, applying seamless mobility in VANETs becomes more challenging. In seamless communications, fast handover is required to decrease the delay in, and guarantee the accessability and service continuity regardless of the vehicle's location and the wireless technology. PMIPv6 and NEMO protocols are the most suitable mobility protocols for the IP-based applications in VANETs, whereas the MIPv6 protocol increases the signaling overhead. With the goal of decreasing the handover delay by early detection of the vehicle's roaming, new adaptations in the MIPv6 protocol have been standardized, such as Fast MIPv6 (FMIPv6) and Hierarchical MIPv6 (HMIPv6). In addition, the PMIPv6 protocol decreases the overhead by introducing local mobility that serves the MN's roaming inside one PMIP domain. However, the signaling overhead increases when the MN moves through multiple domains. Furthermore, the problem of nested NEMO [32], in which signals from nested MRs travel through all parent MR's home agents, increases the seamless communication delay.

1.4.1.2 IPv6-Based Multihop VANETs

In early research, the focus was to support IP mobility for VANET where there is a direct connection between the vehicle and the RSU , i.e., a one-hop connection [33, 34]. However, with the proliferation of Vehicle-to-Vehicle-to-Infrastructure (V2V2I) communications, the need for the multihop VANETs increases. New challenges in applying IP mobility are described below [35]:

- The operations and performance of IP in the current 802.11p/WAVE VANET standard has not been identified clearly yet.
- IP mobility support for the multihop environment adds complexity to VANET scenarios with a short duration multihop connections, given the high-speed VANETs.
- Asymmetric links are difficult to be implemented in V2V2I communications.
- The intermediate vehicle has a lack of motivation to forward generated data to other vehicles.
- Heterogeneity in vehicular networks is not only in wireless technology, but also in type of equipments and applications.

1.4.1.3 Scalability and Efficiency

Vehicular networks could be very large in size, where thousands of vehicles may communicate together. In addition, the frequent changes of the point of attachment for each vehicle coupled with the vehicle high speeds, affect the VANET topology, therefore the IP mobility protocol needs to be scalable and efficient in signalling overhead during handover, in order to support service continuity and seamless communications for vehicles.

1.4.2 Security Challenges

1.4.2.1 Vehicle's Anonymity and Location Privacy

Previous studies have attempted to secure the mobility signaling in mobile networks by focusing on the authentication and integrity problems only [10, 36–38]. Moreover, much recent research work has been done on anonymity and location privacy problems [9, 39, 40]. As mentioned in [41, 42], location privacy threats vary from a simple interference with personal activities and habits, to a more dangerous physical attack after identifying a person and his favorite locations.

1.4.2.2 Authenticating Vehicles in Multihop Communications

In order to support seamless communications, an adaptation for IP mobility management protocol, such as Proxy Mobile IPv6 (PMIP), has been proposed in [43] to provide IP mobility support in an infrastructure-connected multihop vehicular network. In such a multihop PMIP network, an MN uses a Relay Node (RN) for communicating with its Mobile Access Gateway (MAG) (i.e., the point of attachment to the infrastructure). The existing authentication schemes that can authenticate this MN to its MAG , use the RN to only forward the authentication credentials between MN and MAG. However, an extra mutual authentication, between MN and RN, is required to thwart authentication attacks early.

Without that authentication, the mobile node may initiate a denial of service (DoS) attack toward the MAG, or the RN may initiate a fraud attack to mislead the MN. In mobile environments, DoS and fraud attacks can cause service disruptions and financial losses, due to resources' exhaustion and high end-to-end delay [44]. The difficulty of generating a security association between MN and RN, which are arbitrary nodes and have not met each other before, makes proposing an authentication-preserving scheme a challenge. Moreover, if public-key authentication schemes are employed for this MN-RN authentication, they would require a large delay that cannot be tolerated by seamless vehicular communications.

1.4.2.3 Physical-Layer Location Privacy Attacks

In addition to the location privacy mechanisms implemented in the network layer in order to protect MNs while roaming among different networks, we observe that other mechanisms are required to thwart lower-layers location privacy attackers, such as triangulation attackers, that localize the senders from the strength of their transmitted signals. Triangulation attackers can be found when applying NEMO protocol to support public hotspots inside moving vehicles.

Furthermore, existing physical-layer location privacy schemes are limited to power variability [45] that uses different power levels in transmitting packets, obfuscation [46] that confuses attackers by replacing real location information with fake, and adding noise [47] that decreases the accuracy of sender's localization to noise ratio. Those schemes are not appropriate for NEMO-based VANET hotspots. Power variability schemes have been proven as weak solutions, because attackers can easily reveal the original signals' powers. In addition, existing obfuscation schemes disguise the exact user's location by returning to the attacker an expanded area in which the user is located. However, in NEMO-based VANET hotspots, location privacy attackers can get the exact users' locations instead of an obfuscated area with the help of the high-accuracy positioning schemes. Furthermore, adding noise to transmitted signals decreases the overall network performance.

1.5 Summary

In this chapter, the process of integrating IP mobility protocols with vehicular networks has been explained through its three functionalities, namely, vehicular IP address configuration, IP mobility mechanisms, and forwarding IP packets over vehicular networks. With the focus of the IP mobility mechanisms, a review of three different IP mobility protocols, MIPv6, PMIPv6, and NEMO Bs, is presented, and their current security schemes have been introduced. In addition, the challenges of integrating the IP mobility with the vehicular networks, from communication and security perspectives, are explained. specific to security and privacy, we introduce three challenges each of which is related to one of the discussed IP mobility protocols. Those challenges are: (1) anonymity and location privacy in MIPv6 networks, (2) multihop mobile authentication in PMIP, and (3) physical-layer location privacy in NEMO BS. In the subsequent chapters, we study each challenge in more detail and propose solutions for those security problems.

References

1. Zhu, K., Niyato, D., Wang, P., Hossain, E.: In Kim, D.: Mobility and handoff management in vehicular networks: a survey. Wirele. Commun. Mob. Comput. 11(4), 459–476 (2011).
2. Gramaglia, M., Bernardos, C., Soto, I., Calderon, M., Baldessari, R.: Ipv6 address autoconfiguration in geonetworking-enabled vanets: characterization and evaluation of the etsi solution. EURASIP J. Wirel. Commun. Netw. **2012**(1), 19 (2012)
3. Fazio, M., Palazzi, C., Das, S., Gerla, M.: Vehicular address configuration. In: Proceedings of the 1st IEEE Workshop on Automotive Networking and Applications (AutoNet), IEEE GLOBECOM 2006. Citeseer, San Fransisco (2006).
4. Bugti, S., Chunhe, X., Wie, L., Hussain, E.: Cluster based addressing scheme in vanet (canvet stateful addressing approach). In: IEEE 3rd International Conference on Communication Software and Networks (ICCSN 2011), pp. 450–454. IEEE, China (2011).
5. Bugti, S., Chunhe, X., Hussain, E.: Auto-configuration for vanet, integrated with regional code association architecture. In: Proceedings 4th International Conference on Computer Engineering and Technology, pp. 136–140, Bangkok (2012).
6. Mohandas, B., Liscano, R.: Ip address configuration in vanet using centralized dhcp. In: 33rd IEEE Conference on Local Computer Networks, LCN 2008, pp. 608–613. IEEE, Montreal (2008).
7. Baldessari, R., Bernardos, C., Calderon, M.: Geosac-scalable address autoconfiguration for vanet using geographic networking concepts. In: Proceedings of 19th IEEE International Symposium on Personal, Indoor and Mobile Radio Communications, PIMRC 2008, pp. 1–7. IEEE, French Riviera (2008).
8. Perera, E., Sivaraman, V., Seneviratne, A.: Survey on network mobility support. ACM SIGMOBILE Mob. Comput. Commun. Rev. **8**(2), 7–19 (2004)
9. Taha, S., Shen, X.: Anonymous home binding update scheme for mobile ipv6 wireless networking. In: Proceedings of IEEE Global Telecommunications Conference (GLOBECOM 2011), pp. 1–5. IEEE, Houston (2011).
10. Taha, S., Céspedes, S., Shen, X.: EM3A: efficient mutual multi-hop mobile authentication scheme for PMIP networks. In: Proceedings of IEEE ICC 2012, Ottawa (2012).

11. Taha, S., Shen, S.: A link-layer authentication and key agreement scheme for mobile public hotspots in NEMO based VANET. In: Proceedings of Communication and Information System Security Symposium (Globecom12 CISS), Anaheim (2012).
12. Soliman, H., Castelluccia, C., ElMalki, K., Bellier, L.: Hierarchical mobile ipv6 (hmipv6) mobility management. Internet Engineering Task Force, IETF RFC 5380 (2008). www.ietf. org/rfc/rfc5380.txt.
13. Koodli, R.: Mobile ipv6 fast handovers. Internet Engineering Task Force, IETF RFC 5268. www.ietf.org/rfc/rfc5268.txt (2008).
14. Soto, I., Bernardos, C., Calderón, M., Melia, T.: Pmipv6: A network-based localized mobility management solution. Internet Protoc. J. **13**(3), 2–15 (2010)
15. Devarapalli, V., Wakikawa, R., Petrescu, A., Thubert, P.: Network mobility (nemo) basic support protocol. Internet Engineering Task Force, IETF RFC 3963. www.ietf.org/rfc/rfc3963.txt (2005).
16. Perkins, C., et al.: Ip mobility support for ipv4. Internet Engineering Task Force, IETF RFC 3344. www.ietf.org/rfc/rfc3344.txt (2002).
17. Johnson, D., Perkins, C., Arkko, J.: Mobility support in ipv6. Internet Engineering Task Force, IETF RFC 3775. www.ietf.org/rfc/rfc3775.txt (2004).
18. Choi, J., Khaled, Y., Tsukada, M., Ernst, T.: Ipv6 support for vanet with geographical routing. In: Proceedings of 8th International Conference on ITS Telecommunications, ITST 2008, pp. 222–227. Phuket (2008). DOI 10.1109/ITST.2008.4740261.
19. Sandonis, V., Calderon, M., Soto, I., Bernardos, C.: Design and performance evaluation of a pmipv6 solution for geonetworking-based vanets. Ad Hoc Networks (2012). DOI 10.1016/j.adhoc.2012.02.008. http://www.sciencedirect.com/science/article/pii/S157087051200025X
20. Baldessari, R., Festag, A., Abeillé, J.: Nemo meets vanet: a deployability analysis of network mobility in vehicular communication. In: Proceedings of the 7th International Conference on Intelligent Transport Systems Telecommunications (ITST'07), pp. 1–6. IEEE, Sophia Antipolis (2007).
21. Baldessari, R., Zhang, W., Festag, A., Le, L.: A manet-centric solution for the application of nemo in vanet using geographic routing. In: Proceedings of the 4th International Conference on Testbeds and research infrastructures for the development of networks and communities, pp. 1–12. Innsbruck, (2008).
22. Céspedes, S., Shen, X., Lazo, C.: Ip mobility management for vehicular communication networks: challenges and solutions. IEEE Commun. Mag. **49**(5), 187–194 (2011)
23. Faigl, Z., Fazekas, P., Lindskog, S., Brunstrom, A.: Performance analysis of ipsec in mobile ipv6 scenarios. In: 16th Mobile and Wireless Communications Summit IST 2007, pp. 1–5. IEEE, Budapest (2007).
24. Yeh, L., Chang, J., Huang, W., Tsai, Y.: A localized authentication and billing scheme for proxy mobile ipv6 in vanets. In: Proceedings of IEEE ICC 2012, pp. 1008–1013. Ottawa (2012).
25. Binet, D., Martin, A., Gaabab, B.: A proactive authentication integration for the network mobility. In: Proceedings of 3rd International Conference on Wireless and Mobile Communications, ICWMC'07, pp. 53–53. IEEE, Guadeloupe (2007).
26. Forsberg, D., Ohba, Y., Patil, B., Tschofenig, H., Yegin, A.: Protocol for carrying authentication for network access (pana). Internet Engineering Task Force, IETF RFC 5191 (2008). www.ietf.org/rfc/rfc5191.txt.
27. Shi, D., Tang, C.: A fast handoff scheme based on local authentication in mobile network. In: Proceedings of 6th International Conference on ITS Telecommunications, pp. 1025–1028. IEEE, Chengdu (2006).
28. Chuang, M., Lee, J.: Lmam: A lightweight mutual authentication mechanism for network mobility in vehicular networks. In: Proceedings of IEEE Asia-Pacific Services Computing Conference, 2008. APSCC'08, pp. 1611–1616. IEEE, Yilan (2008).
29. Bauer, C.: Network mobility route optimization with certificate-based authentication. In: Proceedings of First International Conference on Ubiquitous and Future Networks, ICUFN 2009, pp. 189–194. IEEE, Hong Kong (2009).

30. Bauer, C.: A secure correspondent router protocol for nemo route optimization. Computer
    Networks. pp. 1–23 (2013). DOI 10.1016/j.comnet.2012.10.021. To appear, http://www.
    sciencedirect.com/science/article/pii/S1389128612004057
31. Bechler, M., Wolf, L.: Mobility management for vehicular ad hoc networks. In: Proceedings of
    61st IEEE Vehicular Technology Conference, VTC, vol. 4, pp. 2294–2298. IEEE, Stockholm
    (2005).
32. Hyung-Jin, L., Dong-Young, L., Tae-Kyung, K., Chung, T.: A model and evaluation of route
    optimization in nested nemo environment. Inst. Electron. Inf. Commun. Eng. IEICE Trans.
    Commun. 88(7), 2765–2776 (2005)
33. Lee, J., Ernst, T., Chilamkurti, N.: Performance analysis of pmipv6-based network mobility
    for intelligent transportation systems. IEEE Trans. Veh. Technol. 61(1), 74–85 (2012)
34. Pack, S., Shen, X., Mark, J., Pan, J.: Mobility management in mobile hotspots with heteroge-
    neous multihop wireless links. IEEE Commun. Mag. 45(9), 106–112 (2007)
35. Umana, S.: Ip mobility support in multi-hop vehicular communications networks. Ph.D. thesis,
    University of Waterloo (2012). http://uwspace.uwaterloo.ca/handle/10012/6889
36. Taleb, T., Letaief, K.: A cooperative diversity based handoff management scheme. IEEE
    Transac. Wirel. Commun. 9(4), 1462–1471 (2010)
37. Kavitha, D., Murthy, K., ul Huq, S.: Security analysis of binding update protocols in route
    optimization of mipv6. In: Proceedings of International Conference on Recent Trends in Infor-
    mation, Telecommunication and Computing (ITC), pp. 44–49. Kochi (2010).
38. Ying, Q., Feng, B.: Authenticated binding update in mobile ipv6 networks. In: Proceedings of
    3rd IEEE International Conference on Computer Science and Information Technology (ICC-
    SIT), pp. 307–311. Chengdu (2010).
39. Lu, R., Lin, X., Zhu, H., Ho, P., Shen, X.: A novel anonymous mutual authentication protocol
    with provable link-layer location privacy. IEEE Trans. Veh. Technol. 58(3), 1454–1466 (2009)
40. Lu, R., Li, X., Luan, T., Liang, X., Shen, X.: Pseudonym changing at social spots: An effective
    strategy for location privacy in vanets. IEEE Trans. Veh. Technol. 61(1), 86–96 (2012)
41. Krontiris, I., Freiling, F., Dimitriou, T.: Location privacy in urban sensing networks: research
    challenges and directions [security and privacy in emerging wireless networks]. IEEE Wirel.
    Commun. 17(5), 30–35 (2010)
42. Whalen, T.: Mobile devices and location privacy: Where do we go from here? IEEE Secur.
    Priv. 9(6), 61–62 (2011)
43. Asefi, M., Cespedes, S., Shen, X., Mark, J.W.: A seamless quality-driven multi-hop data delivery
    scheme for video streaming in urban VANET scenarios. In: Proceedings of IEEE ICC 2011,
    pp. 1–5. Kyoto (2011). DOI 10.1109/icc.2011.5962785.
44. Guo, C., Wang, H.J., Zhu, W.: Smart-phone attacks and defenses. In: Proceedings of 3rd
    Workshop on Hot Topics in Networks, HotNets III. San Diego (2004).
45. El-Badry, R., Sultan, A., Youssef, M.: Hyberloc: providing physical layer location privacy in
    hybrid sensor networks. In: Proceedings of IEEE International Conference on Communications,
    pp. 1–5. IEEE ICC 2010, Cape Town (2010).
46. Ardagna, C., Cremonini, M., Damiani, E., De Capitani di Vimercati, S., Samarati, P.: Loca-
    tion privacy protection through obfuscation-based techniques. In: Proceedings of Data and
    Applications Security XXI, pp. 47–60. Springer, Redondo Beach (2007).
47. El-Badry, R., Youssef, M., Sultan, A.: Hidden anchor: a lightweight approach for physical layer
    location privacy. J. Comput. Syst. Netw. Commu. 2010, 1–12 (2010)

Chapter 2
Anonymity and Location Privacy for Mobile IP Heterogeneous Networks

2.1 Introduction

In transmitting mobile IPv6 binding update messages, both the mobile node's (MN) Home Address (HoA) and Care of Address (CoA) are transmitted as plain-text, hence they can be revealed by network entities and attackers. The mobile node's HoA and CoA represent its identity and its current location respectively. Therefore, revealing an MN's HoA means breaking its anonymity and revealing an MN's CoA means breaking its location privacy. On one hand, some existing anonymity and location privacy schemes [1–4] require intensive computations, hence, they cannot be used in the time-restricted seamless handovers occurring in mobile networks such as VANETs. On the other hand, some other schemes [5, 6] achieve low anonymity and location privacy levels. Therefore, the trade-off between the network performance on one side and the MN's anonymity and location privacy on the other side makes privacy preserving a challenging issue.

In this chapter, based on the onion routing [7] and anonymizer [8], we propose an anonymous and location privacy preserving (ALPP) scheme that consists of two complementary sub-schemes: anonymous home binding update (AHBU) and anonymous return routability (ARR). AHBU and ARR achieve mobile senders' and receivers' privacy by introducing anonymity and location privacy to mobile IPv6 home binding update and return routability control messages, respectively. More specifically, AHBU sends anonymous home binding update messages to the MN's HA, while ARR sends anonymous return routability messages to the MN's CN.

S. Taha and X. Shen, *Secure IP Mobility Management for VANET*,
SpringerBriefs in Computer Science, DOI: 10.1007/978-3-319-01351-0_2,
© The Author(s) 2013

2.2 Preliminaries

2.2.1 Certificate-Less Public Key Cryptography

In the Certificate-less Public Key Cryptography (CL-PKC) [9], a trusted key generator center (KGC) uses a security parameter, K, and runs a setup algorithm to produce two keys, $(s, Param)$. Chosen randomly from $\mathbb{Z}_q^*$, the master key, s, is kept secret at the KGC, whereas the public, $Param = \langle \mathbb{G}_1, \mathbb{G}_2, e, n, P, P_0, H_1, H_2 \rangle$, is transmitted to all network's users. $\mathbb{G}_1$ and $\mathbb{G}_2$ are cyclic groups of a large prime order, q, $\hat{e} : \mathbb{G}_1 \times \mathbb{G}_1 \to \mathbb{G}_2$ is a bilinear pairing function on elliptic curves [10], n is the bit-length of the plaintext, P is $\mathbb{G}_1$'s generator, $P_0 = s \times P$, and $H_1 : \{0, 1\}^* \to \mathbb{G}_1^*$ and $H_2 : \mathbb{G}_2 \to \{0, 1\}^n$ are two hashing functions.

When the KGC receives a request from a user A with identity ID_A, it creates A's partial private key, $D_A = s \times \mathbf{Q}_A$, where $\mathbf{Q}_A = H_1(ID_A)$, and then securely transmits the partial private key, D_A, to A. Therefore, the user A creates its public-private key pair, (P_A, S_A), by using D_A as follows:

$$
\begin{aligned}
& x_A \in_R \mathbb{Z}_q^* \\
& S_A = x_A \times D_A \\
& X_A = x_A \times P \\
& Y_A = x_A \times P_0 = x_A \times s \times P \\
& P_A = \langle X_A, Y_A \rangle
\end{aligned}
\tag{2.1}
$$

Unlike traditional public key infrastructure, Cl-PKC does not require a user's certificate to be authorized from a trusted certificate authority. Therefore, the communication and computation overheads needed for certificate distribution and verification are saved. Algorithm 1 shows the certificate-less encryption of a message m transmitted to a user A. Note that the sender employs only A's identity (ID_A) and public key (P_A) to produce a ciphertext, c. Section 2.6.2 shows that if an adversary changes either A's identity or public key, then the encryption operation results a failure operation ($\perp$) or an incorrect ciphertext. Moreover, to decrypt this ciphertext, $c = \langle u, v \rangle$, A implements only one pairing function to get the message, $m = v \oplus H_2(\hat{e}(S_A, u))$.

In this chapter, we exploit the CL-PKC to generate a shared key between two users, A and B. A transmits its public key along with a random value T_A to B, which responds with its public key P_B and another random number, T_B. $T_A = aP$ and $T_B = bP$, where a and b are randomly chosen by A and B, respectively. Using this transmitted information, both A and B create two keys: K_A is generated by A, and K_B is generated by B, as follows:

Algorithm 1: CL-PK Encryption

Input: m, ID_A, and P_A
Output: Ciphertext c
1 **if** $\hat{e}(X_A, P_0) \neq \hat{e}(Y_A, P)$ **then**
2 | $c = \perp$
3
4 **else**
5 | $Q_A = H_1(ID_A) \in \mathbb{G}_1^*$
6 | $r \in_R \mathbb{Z}_q^*$
7 | $c = \langle rP, m \oplus H_2(\hat{e}(Q_A, Y_A)^r) \rangle$
8 **end**

$$K_A = \hat{e}(Q_B, Y_B)^a . \hat{e}(S_A, T_B) \qquad (2.2)$$

$$K_B = \hat{e}(Q_A, Y_A)^b . \hat{e}(S_B, T_A) \qquad (2.3)$$

Using the pairing function's properties, it can be showed that both keys are identical as follows:

$$\begin{aligned}
K_A &= \hat{e}(Q_B, Y_B)^a . \hat{e}(S_A, T_B) \\
&= \hat{e}(Q_B, x_B s P)^a . \hat{e}(x_A s Q_A, b P) \\
&= \hat{e}(x_B s Q_B, a P) . \hat{e}(Q_A, x_A s P)^b \qquad (2.4) \\
&= \hat{e}(S_B, T_A) . \hat{e}(Q_A, Y_A)^b \\
&= K_B
\end{aligned}$$

2.3 Related Work

2.3.1 Chaum's Mix

The Chaum's mix [11] is the first to introduce the idea of the mix-network, which previous schemes rely on to achieve anonymity and location privacy in mobile IPv6 networks. A mix-network is a group of mix servers that decrypt incoming messages and then retransmit them to the destinations in a different order rather than their incoming order. The goal of this mixing is to hide the sender's identity and locations, and it is employed in the cascaded overlay mix-network-based location privacy schemes [7, 12, 13].

Another Chaum's mix-based scheme, called anonymizer [8], bases on a single trusted proxy to hide user's identity and location information from a Correspondent Node (CN).

Based on the anonymizer, a scheme with eight different levels of anonymity and location privacy is proposed in [14]. This scheme introduces a new entity, called

Information Translating Proxy (ITP), which works as an anonymizer in a mobile IPv6 network. Each mobile node shares a secret key with the ITP, and uses this key to encrypt the home binding update messages at the time of roaming. Instead of sending the binding update messages directly to the mobile node's home agent, the mobile node sends them to the ITP which removes the mobile node's identity information and then forwards these messages to the home agent. Although it presents a practical solution for location privacy, this scheme is susceptible to a single point of failure, because it uses a single trusted anonymizer for all mobile nodes. In our proposed scheme, ALPP, we use the idea of anonymizer, solving the single point of failure problem by changing the anonymizer as the MN moves among visited networks.

2.3.2 Mobile Nodes' Pseudonyms

In [15], an anonymity scheme is introduced which is based on generating groups of pseudonyms and sharing them between the communicating parties. However, this scheme requires a precise synchronization, in such a way that both the sender and the receiver must use the same pseudonym at the same time. In the case of weak synchronization, a collision at the receiver side may happen. Furthermore, a pseudo HoA is randomly generated in [16] to replace the real HoA. Although this pseudo identity achieves the MN's privacy, it may cause a repudiation attack if the MN is a malicious one. A privacy tag is introduced in [17] to hide the MN's HoA from the correspondent node during transmitting the binding update messages from MN to its CN. This privacy tag is a function of the MN's HoA and it prevents the CN from knowing about the MN's roaming.

2.3.3 Mobile IP Location Privacy

In Mobile IPv6 protocol, the mobile IP address represents both an MN's identity and location, Therefore, mobile IP-based networks have location privacy problems. Therefore, in [18], a virtual ID is used to represent MN's identity and hence separate this identity from MN's location. Therefore, additional servers are required to map virtual IDs to MNs' current locations. However, this scheme causes a triangle routing problem, because messages sent from the CN are transmitted to the MN's HA before reaching the intended MN. In [19], a name space is used to represent the MN's identity and a new layer, Host Identity Protocol (HIP), is added to the TCP/IP protocol stack. Supporting mobility and multi-homing is the main goal for the HIP, additionally it provides MN's location privacy service. We argue that HIP protocol is computationally expensive. To initiate a communication between two entities, initiator and responder, HIP uses the entities' public keys for both identification and sharing a secret key between these entities. In addition, the responder transmits a puzzle to the initiator in order to authenticate it, which takes CPU processing time from the initiator to solve the puzzle. By definition, the client puzzle is a difficult problem that requires amount of computations and/or storage to be solved at client side. Therefore, the HIP protocol cannot be used with seamless communications.

Furthermore, a mix-based location privacy scheme is proposed in [4] to achieve anonymity and location privacy for mobile IPv6 binding update control messages. A network of mix servers controlled by a mix center is deployed, and uses (k, n) ElGamal threshold mechanism to decrypt the binding update messages received from the roaming mobile node. This scheme uses the mix-network [11] to hide the MN's location and a pseudo identity to hide the MN's real identity. However, the mix center identifies the mobile node's home address, care of address, home agent, and foreign gateway. Therefore, the mix center can easily violate the mobile node's privacy. Unlike our proposed scheme, the mix-based scheme cannot be used for the time-restricted seamless communications, because it has high routing-delays, especially with a large number of mix servers.

In [5], the Internet Engineering Task Force (IETF) group defines the location privacy problem in the mobile IPv6 networks. The problem definition is divided into two main parts: disclosing the care of address to the correspondent node, and revealing the home address to an eavesdropper. Furthermore, the IETF group published experimental solutions in [6] to solve only the second part of the problem. Those solutions do not address the first part of the location privacy problem, i.e., unveiling the CoA to the correspondent node. Specifically, two schemes are proposed in [6]. The first scheme uses the encrypted home address (EHoA) to conceal the home address from the adversary, while the second uses pseudo identity (PHoA) to hide the home address from the correspondent node. However, EHoA and PHoA schemes achieve only the mobile node's anonymity, and assume that MN's location privacy is implicitly achieved. In our proposed scheme, in addition to the problems defined in [5], we solve two more privacy problems: disclosing the CoA to the HA, and revealing the HoA to the FG. In Sect. 2.6, we show that our proposed sub-schemes, AHBU and ARR, achieve higher anonymity and location privacy levels than those achieved by EHoA and PHoA schemes.

2.4 System Models

2.4.1 Network Model

Figure 2.1 depicts a group of heterogeneous networks, such as VANETs with IEEE 802.11p, WiMaX, and WiFi as heterogeneous access networks. Those heterogeneous networks employ the mobile IPv6 protocol to support the mobile users with mobility services. Therefore, mobile users can receive their communication messages while they are roaming to foreign networks. Each network consists of a number of MNs, such as vehicles, and a set of gateways, such as the Road side units (RSUs) inside VANETs. Each gateway has three functions: (1) to work as a HA for MNs that are

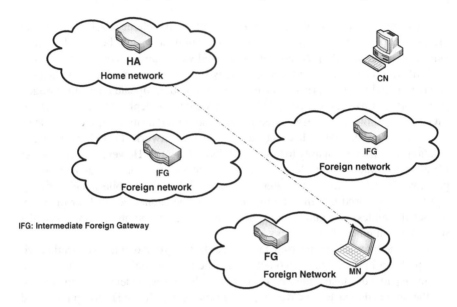

Fig. 2.1 MIPv6 network model

originally located in its network; (2) to work as a foreign gateway (FG) for the visitor MNs; and (3) to work as an intermediate foreign gateway (IFG) for MNs that are neither visitors nor originally located in this gateway's network. Each MN defines its HA, located in its home network, and its FG, located in its current visited network. Moreover, the MN also defines a list of all IFGs, which consists of all gateways that are located in all networks except gateways that are located in both MN's home network and currently visited network.

Each MN maintains a secret key, K_{MN-HA}, that is shared permanently between the MN and its HA, using the IPsec Internet key-exchange protocol [20]. When the MN roams to a foreign network, it sends mobile IPv6 binding update control messages to both its home agent and its CN to inform them about MN's current location. Therefore, any subsequent messages can be directed to the MN's current location (MN's CoA) instead of sending them to its home address.

2.4.2 Threat and Trust Models

We define two kinds of adversaries: external and internal. The external adversary is a passive traffic analysis attacker that analyzes the transmitted packets to acquire useful information about the identities and the locations of the senders. By investigating the time of each transmitted packet, the adversary compares the received and the transmitted packets at each hop, and tracks the packet to know its destination.

On the other hand, the internal adversary is a network entity that intentionally observes MNs' identities and locations. In our model, we consider the HAs, FGs, and CNs entities as internal adversaries. These entities may misuse the observed MNs' privacy information and take malicious actions toward these MNs. Therefore, HAs, FGs, and CNs are prevented from learning MNs' private information. However, these entities need to learn the MN's locations because they help in the MN's mobility management process. To illustrate this contradiction, consider for instance a home binding update message that is transmitted from an MN to its HA. The receiver HA needs to know the MN's identity and current location, and stores this information in the HA's binding cache. Therefore, the HA can forward any subsequent messages destined for the MN's HoA to the current MN's CoA. However, at the same time the HA may maliciously use the MN's information to violate MN's location privacy. To solve this contradiction, we let the internal adversaries learn only part of the mobile nodes' private information. This part is adequate to perform the MN's mobility management process without violating MN's privacy. HAs and CNs are allowed to know mobile nodes' HoAs, however they are unable to learn MNs' care of addresses and foreign gateways. Moreover, the foreign gateway is allowed to know the mobile node's CoAs, and it should not know the MNs' home addresses. All in all, each internal adversary learns a different part of MN's privacy information. Therefore, internal adversaries may collude with each other to know the whole MN's private information.

The HA is consider as a non-colluder with other entities in the network, while all other entities, including the FGs, IFGs, and CNs, are untrusted entities, and they may collude with each other to reveal the mobile node's private information. In addition, there is a trusted third party that generates a group key (K_{group}) for the entire networks. The K_{group} is securely transmitted to all legitimate users in the system.

2.5 Anonymous and Location Privacy Preserving Scheme

In this section, the anonymous and location privacy preserving scheme (ALPP), is proposed, to be used by an MN in seamless handover time, when the MN roams to a foreign network. With the goal of achieving the MN's anonymity and location privacy in this time-restricted seamless handover, ALPP [21] performs three stages: the setup, AHBU, and ARR. The setup stage is required to be implemented, whereas the AHBU and ARR sub-schemes can be independently implemented in the network.

2.5.1 Setup

When an MN roams to a foreign network and becomes under an FG's coverage, it employs the CL-PKC to initiate the setup stage. This FG works as a KGC for

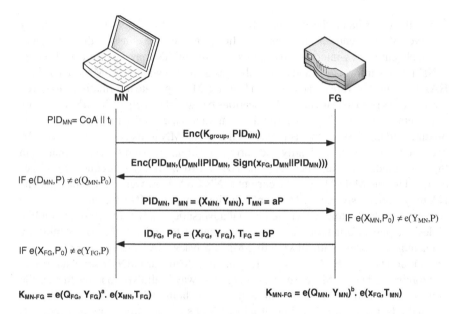

Fig. 2.2 Setup stage

the CL-PKC and periodically sends its identity and its public $Param = \langle \mathbb{G}_1,$ $\mathbb{G}_2, \hat{e}, n, P, P_0, H_1, H_2 \rangle$ to the network's users. Two goal for the setup stage: (1) to mutually authenticate the MN and FG while keeping MN's anonymity; and (2) to establish a shared secret key between those two nodes. The exchanged messages shown in Fig. 2.2 illustrates the mutual authentication as well as the key establishment schemes.

The difficulty of establishing a trust between a two arbitrary nodes, MN and FG, which have not met each other before, is the challenge of this mutual authentication scheme. The following steps summarize the setup stage. The first three steps achieve the anonymous mutual authentication scheme, and the last step achieves the key establishment scheme.

1. The MN creates a pseudo identity, PID_{MN}, by appending a time stamp to its acquired CoA and, i.e., $PID_{MN} = CoA\|t_i$. The MN encrypts the PID_{MN} using the group key, K_{group}, and sends the encrypted message to the FG as follows:

$$FG \leftarrow MN : Enc(K_{group}, PID_{MN})$$

when transmitting this message, the FG guarantees that the MN is a legitimate user. Recall that K_{group} is a secret key shared among all users in the system. The source address of this message is PID_{MN} and the destination address is the FG's address.

2. After authenticating the MN as a legitimate user, the FG creates the MN's partial private key, $D_{MN} = s \times Q_{MN}$, where s and Q_{MN} are defined in Sect. 2.2.1.

Furthermore, the FG signs the D_{MN} along with the PID_{MN} and then transmits them to the MN after encrypting the whole message using the mobile node's pseudo identity, PID_{MN}, as follows:

$$Enc(PID_{MN}, (D_{MN} \| PID_{MN}, Sign(S_{FG}, D_{MN} \| PID_{MN})) \qquad (2.5)$$

It is worth to mention that the MN creates different PID_{MN} at each foreign network. The PID_{MN} involves the CoA, which is related to the FG. Therefore, when the MN communicates with a different FG, its CoA changes and accordingly the PID_{MN} will be changed. This property increases the MN's anonymity level.

3. The MN verifies the FG's signature in the received message and then checks the correctness of the received partial private key, D_{MN}, using the following condition:

$$IF\hat{e}(D_{MN}, P) \neq \hat{e}(Q_{MN}, P_0), wrong D_{MN}$$

The MN then generates its public and private keys, P_{MN} and S_{MN}, using the received partial private key, D_{MN}. When the MN changes its PID_{MN}, the computed public-private key pair will be changed accordingly.

4. Finally, the roaming MN emloys the generated public-private key pair to generate a secret key K_{MN-FG} shared with its FG as illustrated in Algorithm 2:

Algorithm 2: MN-FG shared key establishment

   Input: PID_{MN}, P_{MN}, P_{FG}
   Output: Shared secret key, K_{MN-FG}
1 $FG \leftarrow MN: PID_{MN}, P_{MN} = (X_{MN}, Y_{MN}), T_{MN} = aP$
2 **if** $\hat{e}(X_{MN}, P_0) \neq \hat{e}(Y_{MN}, P)$ **then**
3 $\quad|\quad$ Return illegal MN
4
5 **else**
6 $\quad$ $MN \leftarrow FG$:
7 $\quad$ $P_{FG} = (X_{FG}, Y_{FG}), T_{FG} = bP$
8 $\quad$ **if** $\hat{e}(X_{FG}, P_0) \neq \hat{e}(Y_{FG}, P)$ **then**
9 $\quad\quad|\quad$ Return illegal FG
10
11 $\quad$ **else**
12 $\quad\quad$ at MN: $K_{MN-FG} = \hat{e}(Q_{FG}, Y_{FG})^a.\hat{e}(S_{MN}, T_{FG})$
13 $\quad\quad$ at FG: $K_{MN-FG} = \hat{e}(Q_{MN}, Y_{MN})^b.\hat{e}(S_{FG}, T_{MN})$
14 $\quad$ **end**
15 **end**

2.5.2 Anonymous Home Binding Update Sub-Scheme

The anonymous home binding update sub-scheme (AHBU) [22] is used to add the anonymity and location privacy services to the home binding update control messages. The AHBU sub-scheme involves two main transmitted messages: the binding update, and the binding acknowledgement.

2.5.2.1 Anonymous Binding Update

The roaming MN uses the created shared secret key, K_{MN-FG}, to transmit anonymous binding update messages to its home agent, located in this MN's home network. To send anonymous home binding update message,as depicted in Fig. 2.4, the steps can be summarized as follows:

1. The roaming MN chooses an intermediate foreign gateway, call it home intermediate foreign gateway (HIFG), from the IFGs list. This HIFG is chosen to be any one of the gateways that are located on the shortest path between the MN's current location and MN home-agent's address. To choose this HIFG, the MN first asks its attached FG to broadcast a route request message to request the shortest routing path to its home agent's address. After receiving the route reply message that contains the shortest path, the MN then randomly chooses one gateway from the gateways on the shortest path to be the HIFG. As illustrated later, the MN uses this HIFG as an anonymizer to hide its location from its HA.
2. The mobile node creates an updated version of a binding update (BU) message and encrypts it using the default security protocol, IPsec protocol. The original BU message contains the mobile node's home address, HoA_{MN}, and its current location, CoA_{MN}. However, the updated BU message contains the HoA_{MN} encrypted by the MN-HA shared secret key, K_{MN-HA}, as well as the following additional fields:

   - Source address: Enc(K_{MN-FG}, PID_{MN})
   - Destination address: FG's address
   - Enc(P_{HIFG}, HA's address)
   - HIFG's address

   Note that the source address looks like a wrong IPv6 address format; however, thanks to the setup stage that enables the FG to identify the CoA_{MN}. According to the the setup stage, the FG stores a binding between the encrypted address, Enc(K_{MN-FG}, PID_{MN}), and COA_{MN} as shown in Table 2.1. Figure 2.3 shows an encrypted BU message when it is transmitted from the MN. Using the idea of onion routing, the MN then repeatedly encrypts this binding update message using three different keys: (1) the MN's shared key with its home agent, K_{MN-HA}; (2) the HIFG's public key, P_{HIFG}; and (3) the MN's shared key with the foreign gateway, K_{MN-FG}. The MN then sends this encrypted BU message to its FG.

Table 2.1 Network bindings

Entity	Binding(s)
FG	$PID_{MN} \rightarrow CoA_{MN}$,
	$Enc(K_{MN-FG}, PID_{MN}) \rightarrow CoA_{MN}$
HIFG/CIFG	$Enc(K_{MN-FG}, PID_{MN}) \rightarrow$ FG's address
HA	$HoA_{MN} \rightarrow Enc(K_{MN-FG}, PID_{MN})$,
	$Enc(K_{MN-FG}, PID_{MN}) \rightarrow$ HIFG's address
CN	$HoA_{MN} \rightarrow Enc(K_{MN-FG}, PID_{MN})$,
	$Enc(K_{MN-FG}, PID_{MN}) \rightarrow$ CIFG's address

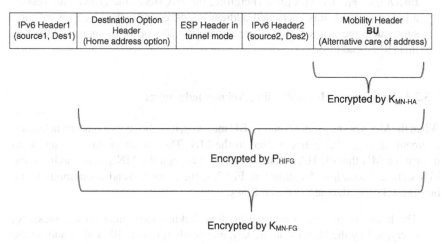

Fig. 2.3 Encrypted binding update message

3. The foreign gateway decrypts the received BU message using its shared key with the MN, K_{MN-FG}, and then sends the decrypted message to the HIFG after adapting the following fields:

   - Source address: FG's address
   - Destination address: HIFG's address
   - $Enc(K_{MN-FG}, PID_{MN})$
   - $Enc(P_{HIFG},$ HA's address)

4. The HIFG stores a binding between the encrypted care of address, $Enc(K_{MN-FG}, PID_{MN})$, and the FG's address. Note that PID_{MN} is a concatenation of MN's CoA and a time stamp t_i. Therefore, for any subsequent messages destined to the encrypted PID_{MN}, the HIFG forwards them to the FG instead. The HIFG also decrypts the encrypted home agent, $Enc(P_{HIFG},$ HA's address), to identify the HA's address and then forwards the BU message to this HA. When the home agent receives the BU message, it contains the following fields:

   - Source address: HIFG's address

- Destination address: HA address
- Encrypted CoA, Enc(K_{MN-FG}, PID_{MN}).
- HoA destination option (HoAD): Enc(K_{MN-HA}, HoA_{MN})

5. The home agent can identify the intended MN from the received HoAD, Enc(K_{MN-HA}, HoA_{MN}). This is because we consider that each HA stores the HoAs of MNs that are under its coverage as well as HoADs. Afterwards, the HA stores a binding between this MN's home address and the encrypted CoA that represents MN's current location. In this binding, the HA cannot identify the MN's current location because it is an encrypted version of the MN's CoA, Enc(K_{MN-FG}, $CoA_{MN}||t_i$). Therefore, the HA stores the HIFG's address as a proxy to reach this encrypted address. Consequently, the HA forwards any subsequent messages destined for the roaming MN or to the encrypted CoA to this HIFG's address.

2.5.2.2 Anonymous Home Binding Acknowledgement

When the MN's home agent receives a BU message, it replies with a binding acknowledgment message that is transmitted to the MN. The reason of this message is to inform the MN that the HA creates a binging between the MN's home address and MN's current location. As shown in Fig. 2.4, the steps to send anonymous home binding acknowledgement are as follows:

1. The home agent creates a home binding Acknowledgement (HBA) message, encrypted by the HIFG's public key, and sends it to the HIFG after adding the following fields:

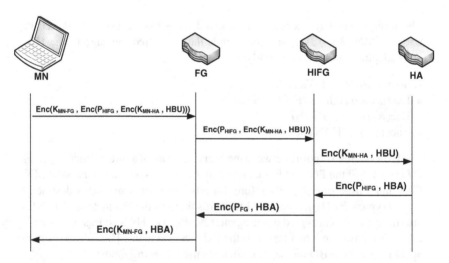

Fig. 2.4 Anonymous home binding update scheme

- Source address: HA's address
- Destination address: HIFG's address
- $Enc(K_{MN-FG}, PID_{MN})$

2. When receiving the HBA message, the HIFG checks its cache memory to identify the corresponding proxy that is attached with the encrypted address, $Enc(K_{MN-FG}, PID_{MN})$. This proxy is the MN's FG; therefore, the HIFG sends the HBA to that FG after encrypting it using the FG's public key and adapting the following fields:

- Source address: HIFG's address
- Destination address: FG's address
- $Enc(K_{MN-FG}, PID_{MN})$

3. The foreign gateway decrypts the received $Enc(K_{MN-FG}, PID_{MN})$ and then forwards the HBA message to the intended mobile node's care of address.

Table 2.1 shows a summary of stored bindings at each network entity.

2.5.3 Anonymous Return Routability Sub-Scheme

In mobile IPv4 networking, a roaming MN communicates with a CN using the reverse tunneling routing method. In this communication, the CN doesn't identify the MN's CoA, which represents MN's current location. Therefore, instead of sending messages directly to the MN's CoA, the CN transmits these messages to the MN's HA, in order to eventually forward the messages to the the MN's CoA. This indirectness in routing achieves mobile node's location privacy. However, the reverse tunneling increases the communication routing delay, and it may lead to a triangle routing problem. This reverse tunneling routing cannot be used for the seamless communications because it increases the handover time and eventually causes a service interruption.

To solve the triangle routing problem, the mobile IPv6 introduces the route optimization routing method. In this routing, the CN uses the shortest routing path to send messages to the roaming MN after identifying this MN's CoA. This path is created by the return routability procedure, which is a group of four messages that are exchanged between the mobile node and the correspondent node. Home Test Init, Care of Test Init, Home Test, and Care of Test are the four messages of the return routability procedure. After successful transmitting these messages, the CN creates a binding between the MN's home address and current location, the MN's CoA, so the CN can directly transmit any subsequent messages to the MN's new location. This direct routing method decreases the routing delay; however, it breaks an MN's location privacy. By monitoring the return routability transmitted messages, the CN as well as an eavesdropper can reveal the MN's anonymity and location privacy.

To add anonymity and location privacy services to the return routability procedure, the anonymous return routability (ARR) sub-scheme is proposed. In the Home Test Init (HTIM) and Home Test (HTM) messages, the mobile node and the correspondent node communicate through the mobile node's home agent (reverse tunneling) to transmit the home-keygen token. Therefore, the AHBU sub-scheme, illustrated in Sect. 2.5.2, can be used to add the MN's privacy for these two messages. HTIM and HTM messages are transmitted from MN to the HA then to the CN. They are similar to BU/BA message because they are also transmitted from MN to HA, so we consider HTIM and HTM messages as BU and BA messages from the transmitted path perspective. Although the messages' formats are different, we can use the same HIFG to transmit HTIM and HTM from MN to HA. Moreover, in the Care of Test Init message (CTIM) and Care of Test message (CTM), the care-of-keygen token is generated through the direct communication between the mobile node and the correspondent node. Therefore, the ARR sub-scheme is proposed to ensure MN's and CN's anonymity and location privacy for both CTIM and CTM messages trans-missions. In the following subsections, two scenarios for the correspondent nodes will be presented: one for a fixed node, and one for a roaming node. In the former scenario, the correspondent node may be a fixed node or a mobile node that is located in its home network at the time of communication with an MN. In the latter scenario, the correspondent node is a mobile node which roams to a foreign network.

2.5.3.1 Fixed Correspondent Node Scenario

In this scenario, we consider that the MN's and the fixed-CN's home addresses are known to each other, However, to achieve location privacy, the MN's current location, CoA_{MN}, is kept unknown to the correspondent node. The ARR sub-scheme consists of two transmitted messages: Care of Test Init, and Care of Test messages. As shown in Fig. 2.5, the CTIM is transmitted from the MN to the CN. The MN first selects an IFG, which we will call correspondent IFG, CIFG. The CIFG is chosen from among those on the shortest path between the MN and CN. The MN then repeatedly encrypts the message using three different keys: (1) the public key of the CN's home

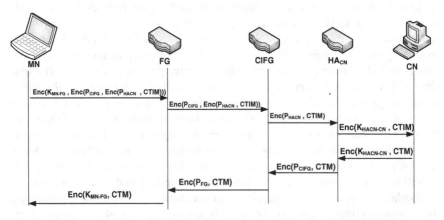

Fig. 2.5 Anonymous return routability, fixed CN

agent, P_{HACN}; (2) the CIFG's public key, P_{CIFG}; and (3) the MN's shared key with its foreign gateway, K_{MN-FG}. The MN then sends the encrypted message to the FG in the foreign network, which in turn forwards the message to the CIFG, and then the message is forwarded to the CN's HA. Finally the CN's HA forwards the message to the intended CN.

When receiving the care of test message, the CN creates a binding between the MN's home address and an encrypted version of the MN's current address, Enc(K_{MN-FG}, PID_{MN}). Furthermore, the CN also stores the address of CIFG as a proxy to reach this encrypted address, Enc(K_{MN-FG},PID$_{MN}$).

The CN then transmits a CTM message to the MN as an acknowledgement for the Care of Test init message. The CN first encrypts the CTM using its shared key with its HA, $K_{HACN-CN}$, and transmits the encrypted message to its home agent. The CN's home agent then encrypts the message with CIFG's public key before transmitting it to the CIFG, which in turn encrypts and transmits the message to the MN's FG. Finally, the MN's FG encrypts the CTM using the shared key with that MN, K_{MN-FG}, and then transmits the encrypted CTM to the MN.

2.5.3.2 Mobile Correspondent Node Scenario

The mobile CN scenario is more difficult than the fixed CN scenario because in this scenario, both the MN and the CN move to two foreign networks. The goal of the ARR sub-scheme here is to achieve MN's and CN's location privacy, which requires hiding the two nodes' current locations from each other. Considering an MN as a mobile sender and a CN as a mobile receiver, we here achieve anonymity and location privacy for both mobile senders and mobile receivers.

We consider that both MN's and CN's home addresses are known to each other. As a mobile node, the CN implements the AHBU scheme, introduced in Sect. 2.5.2, to achieve its anonymity and location privacy towards its home network. To implement the ARR sub-scheme, as shown in Fig. 2.6, the MN sends a CTIM to the correspondent node. First, the CTIM message is sent to the CN's home agent, which discovers that the CN is currently roaming on a foreign network. Therefore, the CN's home agent forwards the message to the CN's CIFG, $CIFG_{CN}$, which in turn transmits the CTIM to the roaming CN.

Going the other way, when the CN sends the CTM to the MN, it is sent directly to the MN's CIFG, $CIFG_{MN}$. The CTM message is not transmitted to the CN's HA because $CIFG_{CN}$ already knows the $CIFG_{MN}$'s address, and thus does not need to ask CN's HA about the $CIFG_{MN}$'s address. Therefore, the length of the CTM routing path is shorter than the length of the CTIM routing path, so the CTM routing path is used for data transmission between a roaming MN and a CN.

The worst case is when the CN and the MN move to the same foreign network. In this case, the two nodes select either the same or different FGs. If both nodes choose the same FG, then only this FG realizes that they are in the same network. Therefore, the FG delivers the messages between the MN and CN without forwarding them to the corresponding CIFGs. If the two nodes choose two different FGs in the

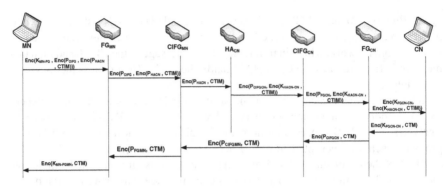

Fig. 2.6 Anonymous return routability, mobile CN

same foreign networks, the MN-CN routing path goes through the corresponding CIFGs and this leads to high routing delay.

2.6 Privacy and Security Analysis

2.6.1 Privacy Analysis

In our network, the mobile node's HoA and CoA represent its identity and its current location respectively. Therefore, violating an MN's HoA means breaking its anonymity, and violating an MN's CoA means breaking its location privacy.

As in [23], we employ the entropy model to measure the degree of anonymity for both our proposed scheme and the mix-based scheme [4]. The degree of anonymity, d, is measured by the following equation:

$$d = 1 - \frac{H_M - H(X)}{H_M} = \frac{H(X)}{H_M} \tag{2.6}$$

$H(X)$ is the entropy of the network, which measures the amount of information that an attacker knows about the identity of message's sender. H_M is the maximum entropy of the network. Therefore, the degree of anonymity for ALPP scheme can be measured as follows:

$$H(X) = \sum_{i=0}^{n} [p_i \log \frac{1}{p_i}] = \log n$$

$$H_M = \sum_{i=0}^{L.n} [p_i \log \frac{1}{p_i}] = \log(L.n) \qquad (2.7)$$

$$d = \frac{\log n}{\log(L.n)}$$

where p_i is the probability that a node i is the sender of a message, n is the number of nodes in the home network, and L is the number of networks in the system.

Similarly, the degree of anonymity for the mix-based scheme can be computed as follows:

$$d = \begin{cases} \frac{\log m}{\log(L.n)} & K = 1, \\ \frac{\log(K.m)}{\log(L.n)} & K > 1. \end{cases}$$

where K is the number of mix servers, L is the number of networks in the system, and m is the number of messages that are mixed together at each mix server. The number of mixed messages is an indicator for the number of senders, because in the mix-based scheme, each sender sends one message at a time to the mix server. Therefore, m also represents number of senders in the network.

Figure 2.7 depicts the degree of anonymity for our scheme at different values of L and for the mix-based scheme with one mix server ($K = 1$). ALPP's degree of anonymity increases as the number of nodes in the home network increases, but it decreases as the number of networks in the system increases. On the other hand, the degree of anonymity for the mix-based scheme increases as the number of senders increases. For the mix-based scheme, we fix the number of users in one network to be 1,000 users. Therefore, for $L = 10$, the total number of users is 10,000.

Compared to our proposed scheme, ALPP, the mix-based scheme with one mix server achieves a lower level of anonymity when number of senders is below 1,000. Increasing the number of senders in the mix-based scheme causes a high delay, as will be shown later. Moreover, increasing the number of mix servers leads to increasing the level of anonymity, but also increases the network delay. This trade-off prevents the mix-based scheme from being used for seamless communications, which require low routing delay to achieve service continuity.

To illustrate the impact of delays on the mix-based scheme, Fig. 2.8 depicts the delay of the scheme multiplied by the achieved degree of anonymity. Assuming 2 ms for the mix server to send and receive a message, the mix-based scheme with one mix server requires around 1.2 s to serve 1,000 senders. This delay increases to around 5 s as the number of mix servers increases. To achieve higher anonymity using one mix server, the number of senders, m, that send messages to this mix server should be increased. For one mix server, m ranges from zero to the total number of users in the system ($L.n$). However, the network delay increases as m increases, because the mix server needs to wait until receiving all messages from all senders, then mixes and retransmits them. Alternatively, the anonymity level can be increased when the

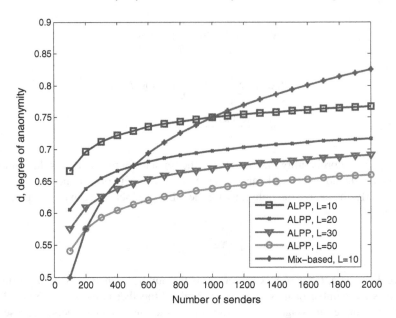

Fig. 2.7 Degree of anonymity

number of mix servers, K, increases. In this case, the number of senders, m, is limited to $0 \leq m \leq \frac{L.n}{K}$. However, the network delay also increases when the number of mix servers increases, because these mix servers work sequentially with each other. As a conclusion, in the mix-based scheme, there is a trade-off between the achieved anonymity level and the network delays.

On the other hand, Fig. 2.9 shows the delay of the ALPP scheme multiplied by its degree of anonymity. Compared to the mix-based scheme, our scheme has a delay of 1.5 ms to serve 1,000 users, which is 99 % less than the mix-based scheme's delay.

The proposed ALPP scheme ensures sender's and receiver's location privacy by hiding their care of addresses from both the home agent and correspondent node. In our network, the care of address and the foreign gateway represent a node's location information. The mobile node's home agent cannot determine the mobile node's

Table 2.2 Mobile node's information knowledge

	HoA	CoA	HA	FG	CN	IFG
MN	K	K	K	K	K	K
HA	K	U	K	U	K	K
FG	U	K	K	K	U	K
IFG	U	U	K	K	K	K
CN	K	U	K	U	K	K
Adversary	U	U	U	U	U	K

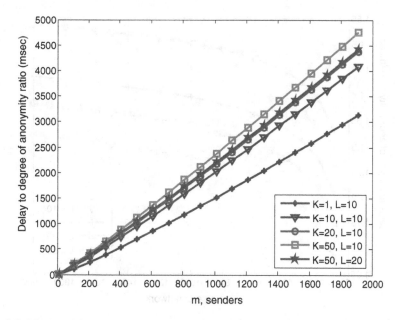

Fig. 2.8 Mix-based delay to degree of anonymity ratio

care of address, because it receives an encrypted address, $\text{Enc}(K_{MN-FG}, PID_{MN})$, instead of a plain-text address. Moreover, the home agent does not communicate directly with the foreign gateway, they communicate through a proxy, IFG. Therefore, the home agent cannot identify the MN's FG.

Table 2.2 shows the mobile node's information that each entity in the network can acquire. In the table, the header column represents network entities, the header row represents MN's information, K means that the network entity knows the information part, and U means that the information is unknown to the network entity. As shown in the table, no network entity except the MN itself can identify this node's location, CoA. Two columns in the table represent MN's location information, the CoA and FG columns. The CoA column show that no entity knows the MN's CoA except this MN and its FG. Although the FG identify the MN's location, it does not identify this MN because the MN communicates with the FG by means of a pseudo identity instead of its real identity. In addition, the FG column shows that the MN, FG, and IFG know MN's FG. The IFG only specifies MN's FG, but it does not identify the MN itself.

2.6.2 Security Analysis

The security of the ALPP scheme is based on the security of the proposed key establishment scheme that is illustrated in Algorithm 2. Moreover, the security of

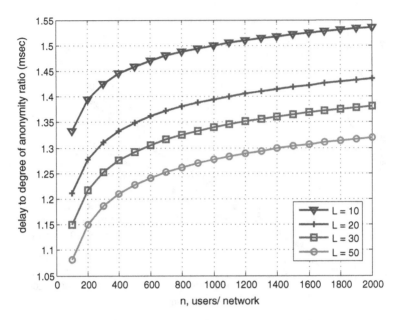

Fig. 2.9 ALPP's delay to degree of anonymity ratio

the key establishment scheme is based on the hardness of the elliptic curve discrete logarithm problem (ECDLP). In [24] it is proved that ECDLP can be solved in at least sub-exponential time. ECDLP is a hard problem since no polynomial time algorithm can solve it.

Definition 1 *The Elliptic Curve Discrete Logarithm Problem (ECDLP):*
  Given P and xP as two points on elliptic curve E, find x where $x \in \mathbb{Z}_q^*$.

Theorem 1 *In Algorithm 2, under the assumption that an attacker knows the MN's private key, S_{MN}, the attacker is still unable to create the shared key K_{MN-FG}.*

Proof. To create a valid K_{MN-FG}, the attacker needs to compute the following pairing functions:

$$K_{MN-FG} = \hat{e}(\mathbf{Q}_{FG}, Y_{FG})^a . \hat{e}(S_{MN}, T_{FG})$$

Since the attacker knows S_{MN}, it easily computes $\hat{e}(S_{MN}, T_{FG})$. However, to compute $\hat{e}(\mathbf{Q}_{FG}, Y_{FG})^a$, the attacker needs to know the value of a. But the attacker knows only P and $T_{MN} = aP$. Thus this problem is equivalent to ECDLP. Since ECDLP is a hard problem, the attacker cannot create a valid K_{MN-FG} in a polynomial time.

2.6.2.1 The Traffic Analysis Attack

The traffic analysis attacker attempts to capture a group of the transmitted packets and analyze them in order to learn the identity and the location of the mobile node. The identity of the mobile node, which is represented by its home address, is transmitted in an encrypted form. Therefore, the traffic analysis attacker cannot learn the true identity of the mobile node. Moreover, we use onion routing to prevent the attacker from correlating the input and output messages at a specific hop. For example, the binding update messages that are transmitted from an MN are repeatedly encrypted by three different keys: the shared key with the home agent, the intermediate foreign gateway's public key, and the shared key with the foreign gateway. When the foreign gateway receives these messages, it decrypts them using the shared key with the mobile node, and then retransmits the decrypted messages to the IFG. These decrypted messages are indeed messages encrypted by the remaining two keys. Therefore, at each hop, the messages are decrypted by one key then retransmitted to the next hop. Consequently, the attacker cannot identify the mobile node's movements.

2.6.2.2 The Collusion Attack

The collusion attack may be triggered among the foreign gateways, the intermediate foreign gateways, or the correspondent nodes. When our proposed schemes are used, a collusion attacker gains no information about the mobile node's identity and locations.

If the foreign gateways collude with each other, they would not learn the identity of the mobile node. In the setup stage, the mobile node uses a pseudo identity, $PID_{MN} = CoA\|t_i$, to identify itself to the FG. The MN's CoA, which is used to create the PID_{MN}, changes as the mobile node chooses different foreign gateways; hence MN's PID_{MN} also changes. Therefore, each foreign gateway identifies only one PID_{MN} of the MN's pseudo identities. It is thus not possible for the FGs to link all the care of addresses to the same MN.

Moreover, the collusion of the intermediate foreign gateways reveals nothing about MN's privacy, because they do not directly communicate with this MN. In our network, IFGs only communicate directly with the home agent and the foreign gateway. The IFG receives an MN's encrypted CoA, which represents the MN's location. Again when it roams among different foreign networks, the MN acquires different care of addresses and encrypts them by different keys. Therefore, if IFGs collude, they cannot link all encrypted CoAs to the same MN. Collusion among the FGs and the IFGs can reveal the mobile node's home agent, but this knowledge of the mobile node's home agent does not break the MN's privacy, because we argue that there are at least two nodes in the home network. Therefore, the probability of identifying the mobile node is:

$$P(MN) = \frac{1}{n}, n \geq 2 \tag{2.8}$$

where n is the number of nodes in the home network. Thus, with a large number of nodes located in the MN's home network, the probability of identifying the MN after identifying its network is negligible.

2.6.2.3 The Replay Attack

In the setup stage, an attacker may send a previously transmitted pseudo identity to the foreign gateway in order to deceive the foreign gateway and learn the MN's partial private key, D_{MN}. In our proposed schemes, the MN's pseudo identity, $PID_{MN} = CoA \| t_i$, is created by concatenating MN's CoA with the time stamp. The time stamp prevents the attacker from repeating transmission of previous messages. However, any legitimate user who knows the group key can decrypt the message, change the time stamp, and then resend the message again. From Theorem **??**, we prove that even if a legitimate user succeeds in learning the MN's secret key, this user is still unable to create a valid shared key, K_{MN-FG}.

2.6.2.4 The MITM Attack

A man-in-the-middle (MITM) attacker may change either MN's identity, PID_{MN}, or public key, P_{MN}, to create a fake session with the FG. We prove by Theorem 2 that if either PID_{MN} or P_{MN} is changed in the middle of transmission, then the key generation algorithm returns "illegal MN".

Theorem 2 *If either PID_{MN} or $P_{MN} = (X_{MN}, Y_{MN})$ is changed by an attacker, then Algorithm 2 returns "illegal MN".*

Proof. **Case 1: If PID_{MN} is changed to $PI\grave{D}_{MN}$,then from Theorem 1, the attacker cannot create $K_{MN-FG} = \hat{e}(Q_{FG}, Y_{FG})^a.\hat{e}(S_{MN}, T_{FG})$ because the attacker does not know the values of a and S_{MN}. Thus the attacker is an illegal MN.**

Case 2: If P_{MN} is changed to $P\grave{M}N = (X\grave{M}N, Y\grave{M}N)$, then the condition at line 2 of Algorithm 2 is satisfied. This means $\hat{e}(X\grave{M}N, P_0) \neq \hat{e}(Y\grave{M}N, P)$, and Algorithm 2 returns "illegal MN".

In addition, an MITM attacker may send a fake partial private key, D_{MN}, to the MN in the setup stage. This case also happens if the FG is a malicious node and wants to mislead the MN. The result of this attack leads to an interruption of the MN's IP session. In our proposed schemes, however, the MN authenticates the FG by verifying its signature as illustrated in the setup stage- further, the MN also checks the correctness of the partial private key that is received from the FG, using the following condition:

$$IF \hat{e}(D_{MN}, P) \neq \hat{e}(Q_{MN}, P_0), wrong D_{MN}$$

We can show that for a correct D_{MN}, the two pairing functions are identical, as follows:

$$\hat{e}(D_{MN}, P) = \hat{e}(s \times \mathbf{Q}_{MN}, P)$$
$$= \hat{e}(\mathbf{Q}_{MN}, s \times P) \qquad (2.9)$$
$$= \hat{e}(\mathbf{Q}_{MN}, P_0)$$

2.7 Performance Evaluation

2.7.1 Computation and Communication Overhead

Tables 2.3 and 2.4 show the computation and communication overheads of the proposed sub-schemes, AHBU and ARR, compared to those of the mix-based scheme [4] with one mix server, and the EHoA and PHoA schemes [6]. In addition, we use Crypto++ benchmarks [25] to measure the computation time at the mobile node's side, as shown in Fig. 2.10. We use the ElGamal encryption mechanism for public key encryption operations, and the AES scheme for symmetric encryptions. According to Crypto++ benchmarks, the modulus and exponent sizes used for ElGamal encryption are 2,048 and 226 bits, respectively. Therefore, in the tables, T_{ELG} represents the time needed for ElGamal encryption operation, T_{Sym} represents the time needed for AES encryption or decryption, T_{pid} and T_{prf} represent the time needed to construct a pseudonym and to generate a random number, and $T_{EHOA-reg}$ and $T_{PHOA-reg}$ represent time needed for registering the encrypted and the pseudo home addresses. For computation overhead, $B_{Signalling}$ represents bytes needed to send the control information, while $B_{EHoA-reg}$ and $B_{PHoA-reg}$ represent bytes needed to send a PHoA and EHoA registration messages.

In Table 2.3, our AHBU's computation overhead is smaller than the mix-based scheme's overhead by 66 %. The mix-based scheme requires three public key encryption operations while the AHBU scheme requires only one public key operation.

Table 2.4 shows that the ARR sub-scheme is the second least time-consuming after the mix-based. In EHoA and PHoA schemes, an MN needs first to register the encrypted home address and pseudo home address before using them. Considering 5 ms for one Round Trip Time (RTT) between the MN and its home agent, the computation overheads of EHoA and PHoA schemes are much higher than that of the ARR sub-scheme. ARR's computation overhead is smaller than the overhead of EHoA

Table 2.3 AHBU computation and communication overheads

	Computation	Communication
Mix-based	$3T_{ElG} + 2T_{Sym}$ $+ 2T_{prf} + T_{pid}$	$B_{signaling}$
AHBU	$T_{ElG} + 3T_{sym}$	$B_{signaling}$

Table 2.4 ARR computation and communication overhead

	Computation	Communication
Mix-based	$3T_{sym} +$ $T_{Hash} + 2T_{pid}$	$B_{Signaling}$
EHOA	$3T_{sym} + T_{EHoA-reg}$	$B_{EHoA-reg}$
PHOA	$T_{Pid} + 2T_{PHoA-reg}$	$B_{PHoA-reg}$
ARR	$2T_{sym} + 2T_{ElG}$	$B_{Signaling}$

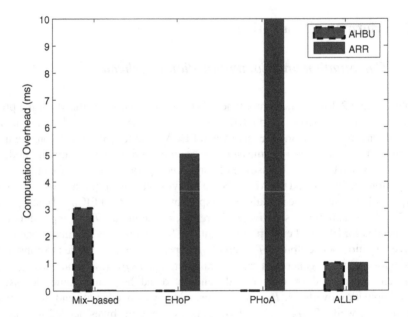

Fig. 2.10 ALPP computation overhead

and PHoA by 79 and 89 %, respectively. Figure 2.10 shows the time consumption for the ALPP scheme compared to other schemes.

The measured AHBU and ARR computation overheads do not include the time required for the setup stage, T_{setup}, because this time is only needed once per MN's stay on a given foreign network. If an MN sends many home binding update messages from the same foreign network, then only one T_{setup} is required. The setup time can be measured as follow:

$$T_{setup} = 2T_{Sym} + T_{verification} + 3T_{pairing} \qquad (2.10)$$

Considering the AES scheme for T_{sym}, and the RSA scheme for signature verification time, $T_{verification}$, the estimated time needed for T_{setup}, is around 120 ms. To measure the pairing time, $T_{pairing}$, we consider a 2.93 GHz processor with the Tate pairing in [25] and get 6.83 ms for each pairing function. Following the Crypto++

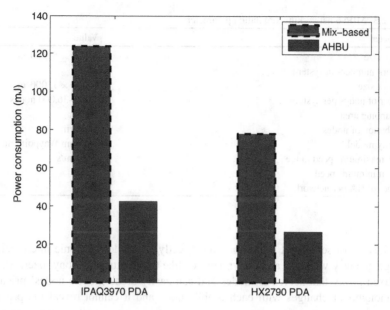

Fig. 2.11 AHBU power consumption

benchmarks, the used Elliptic Curve field size is $GF(2^n)$ where operations are implemented using trinomial basis. We also use 17 as the RSA scheme public exponent.

2.7.2 Power Consumption

Aiming to compute the energy consumed at MN, we follow the energy costs of cryptographic algorithms that are proposed in [26] for two different PDAs, a Compaq iPAQ3970 and an HP Hx2790. As shown in Fig. 2.11, compared to the Mix-based scheme, the AHBU sub-scheme has the lowest energy consumption for both PDA types. AHBU achieves energy reductions of 65.66 and 66 % when using Compaq iPAQ3970 and HP Hx2790, respectively. This is due to using only one public key operation, while the mix-based scheme uses three. According to [26], one public key scheme requires 40.87 and 25.87.17 mJ to encrypt a message in iPQ3970 and Hx2790, respectively.

2.7.3 Simulation Results

Based on the anonymizer scheme [8], we have proposed a new method of routing in which the transmitted binding update message is sent to an intermediate node,

Table 2.5 MIPv6 networking simulation parameters

Parameter	Value
System size	$5,500 \times 5,500$ m
Network numbers in system	36
Network size	$1,000 \times 1,000$ m
Number of nodes per system	1,000–36,000 nodes
Overlapping area	100 m
Distribution of nodes	Uniform
Mobility model	Random Waypoint model
Nodes maximum speed range	2–20 m/s
Nodes minimum speed	0 m/s
Number of HA per network	one

IFG, instead of sending it to the receiver directly. The selected home intermediate foreign gateway works as an anonymizer. Unlike the traditional anonymizer, which is a fixed proxy that serves all nodes and can easily break mobile nodes' privacy, our anonymizer changes with each mobile node and it cannot reveal the privacy information.

We develop a simulator to compare the effect of the updated routing method used by both AHBU and ARR sub-schemes with that of the original routing method, which does not ensure privacy for any MN.

Two kinds of mobile nodes are defined in our simulator. The first type, called the successful node, is the node that succeeds in finding an intermediate foreign gateway on the shortest path between the communicating parties. The second type, called the failed node, is the node that moves to a neighbor network, so the shortest path length is only one hop, from HA to FG. Therefore, the failed node cannot find an intermediate gateway on the shortest path.

We consider 351 simulation runs, where the number of nodes in the system increases from 1,000 nodes in the first run, to 36,000 nodes in the last. We consider a large number of nodes in order to check the scalability of our proposed scheme. At each run, the maximum node speed ranges from 2 to 20 m/s. The time interval between each run is 10 min. We use the Bellman-Ford routing algorithm for message routing among gateways. Table 2.5 shows the full simulation parameters.

Figure 2.12 shows the routing delays of the proposed sub-schemes, compared with the home binding update (HBU) scheme and the triangle routing that is used by the mobile IPv4 protocol. HBU and triangle routing schemes are used as lower and upper references, respectively. In this figure, we measure the routing delay for a highly dense networking, 36,000 nodes density.

As shown in the figure, the proposed sub-schemes (AHBU, ARR sub-scheme with fixed CN scenario, and ARR sub-scheme with mobile CN scenario), have very similar routing delays to that of the HBU. The HBU scheme does not apply any anonymity or location privacy services. This result indicates the ability of using our proposed schemes with scalable networks and real-time applications in which the

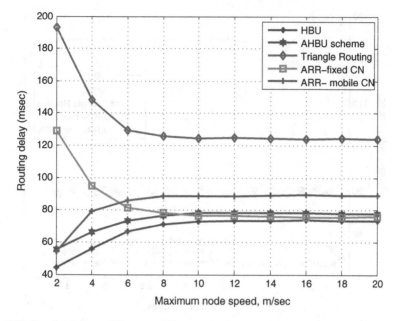

Fig. 2.12 Routing delay at different mobility speeds

routing delay is an important factor. The reported difference in the routing delays between our proposed schemes and the HBU results from the failed nodes. In our simulation, the failed nodes do not apply our updated routing method, since there is no IFG on the shortest path. An alternative solution is that the failed nodes can select any IFG at an adjacent network. In this case, routing delay values may depend on the network traffic, because the adjacent network is not located on the shortest path between the two communicating parties.

We also notice that the routing delay of the triangle routing method is larger than our schemes' delays. The triangle routing method ensures an MN's location privacy, but its high delay prevents its use for seamless communications. Our sub-schemes' routing delays are smaller than the triangle routing delay by an average of 32 %.

Figure 2.13 shows the network routing delay for different network capacities at high node mobility, 20 m/s. It can be seen that the number of nodes in the network does not have a significant impact on the routing delay, but the nodes' mobility speed has a large impact on this delay. Compared to the HBU scheme, our proposed sub-schemes, the AHBU, ARR-fixed CN, and ARR-mobile CN schemes, increase the routing delays by 2.7, 4, and 20 % respectively. On the other hand, compared to the triangle routing scheme, our sub-schemes decrease the routing delays by 42, 43, and 30 % respectively.

Figure 2.14 shows the number of successful and failed nodes with a 36,000-node system and at different node speeds. It can be seen that the number of failed and successful nodes depends on the speed of the node. With mobility speeds below

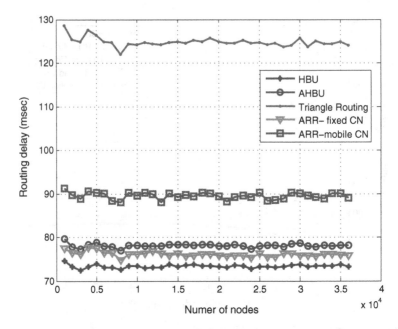

Fig. 2.13 Routing delay with different network capacity

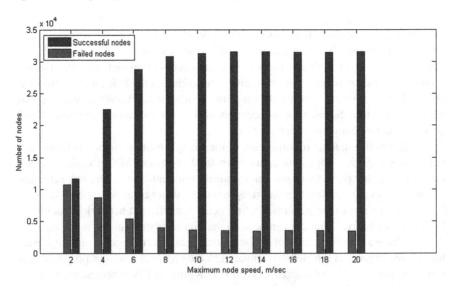

Fig. 2.14 Successful and failed nodes at mobility speeds

Table 2.6 95% confidence interval of AHBU sub-scheme

Successful nodes

Density	Mobility	Mean	St.dev	CI
low	low	868	44.26	[781, 955]
low	high	876	10.55	[855.7, 897.1]
high	low	31,273	1,516.7	[28,300, 34,246]
high	high	31,465	112.88	[31,244, 31,686]

Average delay

low	low	73.22	1.58	[70.12, 76.32]
low	high	73.81	0.99	[71.86, 75.76]
high	low	73.63	1.32	[71.04, 76.22]
high	high	73.36	0.18	[73, 73.72]

8 m/s, the number of the successful nodes increases as the nodes' speeds increase. However, with mobility speeds above 8 m/s, the numbers of successful and failed nodes are fixed. This result confirms that our schemes are more appropriate to be used in high mobility environments.

Additionally, we obtain the 95 % confidence intervals (CIs) for both the successful nodes numbers' and the average routing delay. Table 2.6 shows the CIs with different system densities and mobility speeds, in which we consider low density as 1,000 nodes and high density as 36,000 nodes. Similarly, we consider low mobility as 2 m/s and high mobility as 20 m/s.

2.8 Summary

In this chapter, based on the onion routing, anonymizer, and certificate-less public key cryptography, we have proposed an anonymous and location privacy preserving scheme (ALPP) to be employed in MIPv6 control signalling for heterogeneous networks. In addition, we have introduced a mutual authentication scheme as well as a key establishment scheme to be used among mobile nodes and foreign gateways in visiting networks. With a large number of senders in our system, the degree of anonymity in our scheme ends up between 60 and 85%, whereas the mix-based scheme encounters a high delay while increasing the degree of anonymity. In addition, using extensive simulations, we show that our proposed sub-schemes decrease the routing delays by 42% for the AHBU sub-scheme, 43% for ARR-fixed correspondent node, and 30% for ARR-mobile node, compared to the triangle routing scheme.

References

1. Wiangsripanawan, R., Safavi-Naini, R., Susilo, W.: Location privacy in mobile ip. In: Proceedings of 7th IEEE Malaysia International Conference on Communication., Proceedings of 13th IEEE International Conference on. Networks, vol. 2, pp. 1120–1125. (2005)
2. Fasbender, A., Kesdogan, D., Kubitz, O.: Analysis of security and privacy in mobile ip. In: Proceedings of 4th International Conference on Telecommunication Systems, Modeling and Analysis, pp. 1–17. Nashville (1996)
3. Escudero-Pascual, A., Hdenfalk, M., Heselius, P.: Flying freedom: location privacy in mobile internetworking. In: Proceedings of INET2001 Conference, pp. 1–7. Stockholmsmässan - Stockholm, Sweden (2001)
4. Jiang, J., He, C., ge Jiang, L.: A novel mix-based location privacy mechanism in mobile ipv6. Comput. Secur. **24**(8), 629–641 (2005)
5. Koodli, R.: Ip address location privacy and mobile ipv6: problem statement. Internet Engineering Task Force, IETF RFC 4882 (2007) http://www.ietf.org/rfc/rfc4882.txt.
6. Koodli, R., Zhao, F., Qiu, Y.: Mobile ipv6 location privacy solutions. Internet Engineering Task Force, IETF RFC 5726 (2010) http://www.ietf.org/rfc/rfc5726.txt.
7. Goldschlag, D., Reed, M., Syverson, P.: Onion routing for anonymous and private internet connections. Commun. ACM **42**(2), 39–41 (1999)
8. The Anonymizer http://www.anonymizer.com/
9. Al-riyami, S.S., Paterson, K.G., Holloway, R.: Certificateless public key cryptography. Adv. Cryptology-ASIACRYPT **2894**, 452–473 (2003)
10. Lee, E., Lee, H.S., Park, C.M.: Efficient and generalized pairing computation on abelian varieties. IEEE Trans. Inf. Theory **55**(4), 1793–1803 (2009) [10.1109/TIT.2009.2013048]
11. Chaum, D.: Untraceable electronic mail, return addresses, and digital pseudonyms. Commun. ACM **24**(2), 84–90 (1981)
12. Dingledine, R., Mathewson, N., Syverson, P.: Tor: the second-generation onion router. In: Proceedings of the 13th Conference on USENIX Security Symposium, vol. 13, pp. 21–37. San Diego (2004)
13. Reiter, M., Rubin, A.: Crowds: anonymity for web transactions. ACM Trans. Inf. Syst. Secur. (TISSEC) **1**(1), 66–92 (1998)
14. Choi, S., Kim, K., Kim, B.: Practical solution for location privacy in mobile ipv6. In: Chae, K.J., Yung, M. (eds.) Information Security Applications, Lecture Notes in Computer Science, vol. 2908, pp. 1965–1976. Springer, Heidelberg (2004)
15. Arkko, J., Nikander, P., Näslund, M.: Enhancing privacy with shared pseudo random sequences. In: C.B.M.J.R.M. Christianson, B. (ed.) Security Protocols. Lecture Notes in Computer Science, vol. 4631, pp. 187–196. (2007)
16. Deng, R., Qiu, Y., Zhou, J., Bao, F.: Protecting location information of mobile nodes in mobile ipv6. In: Proceedings of First International Conference on Communications and Networking, pp. 1–7. Beijing, China (2006)
17. Koodli, R., Devarapalli, V., Flinck, H., Perkins, C.: Short paper: location privacy with ip mobility. In: Proceedings of First International Conference on Security and Privacy for Emerging Areas in Communications Networks, pp. 222–224. Athens, Greece (2005)
18. So-In, C., Jain, R., Paul, S., Pan, J.: Virtual id: a technique for mobility, multi-homing, and location privacy in next generation wireless networks. In: Proceedings of 7th IEEE Consumer Communications and Networking Conference (CCNC), pp. 1–5. IEEE, Las Vegas (2010)
19. Nikander, P., Gurtov, A., Henderson, T.: Host identity protocol (hip): connectivity, mobility, multi-homing, security, and privacy over ipv4 and ipv6 networks. IEEE Commun. Surv. Tutorials **12**(2), 186–204 (2010)
20. Kaufman, C., Hoffman, P., Nir, Y., Eronen, P.: Internet key exchange protocol version 2 (ikev2). Internet Engineering Task Force, IETF RFC 5996 (2010) http://www.ietf.org/rfc/rfc5996.txt.
21. Taha, S., Shen, X.S.: Alpp anonymous and location privacy preserving scheme for mobile ipv6 heterogeneous networks. Secur. Commun. Netw. **6**(4), 401–419 (2013). 10.1002/sec.625. http://dx.doi.org/10.1002/sec.625

22. Taha, S., Shen, X.: Anonymous home binding update scheme for mobile ipv6 wireless networking. In: Proceedings of IEEE Global Telecommunications Conference (GLOBECOM 2011), pp. 1–5. IEEE, Houston (2011)
23. Diaz, C., Seys, S., Claessens, J., Preneel, B.: Towards measuring anonymity. In: Proceedings of Privacy Enhancing Technologies, PET2003, pp. 184–188. Springer, Dresden (2003)
24. Diem, C.: On the discrete logarithm problem in elliptic curves. technical report (2009)
25. Dai, W.: Crypto++ 5.6. 0 benchmarks http://www.cryptopp.com/benchmarks.html
26. Rifa-Pous, H., Herrera-Joancomarti, J.: Computational and energy costs of cryptographic algorithms on handheld devices. Future Internet **3**(1), 31–48 (2011)

Chapter 3
Multihop Mobile Authentication for PMIP Networks

3.1 Introduction

An important requirement for current mobile wireless networks, such as VANETs, is that they be able to provide ubiquitous and seamless IP communications in a secure way. Moreover, these networks are envisioned to support multihop communications, in which intermediate nodes help to relay packets between two peers in the network. Therefore, in infrastructure-connected multihop mobile networks, the connection from the mobile node (MN) to the point of attachment may traverse multiple hops (Fig. 3.1).

The reasons for relaying packets in infrastructure-connected mobile networks are twofold: (1) direct connection to the infrastructure may not always be available, thus, by using relayed communications, the network coverage can be extended, and its throughput and capacity can also be increased [1]; and (2) Relay Nodes (RNs) may benefit from offering their services as temporary relays; different cooperation incentive schemes have shown that it is in the best interest of each node to participate in multihop packet forwarding and earn credits that reward them per forwarded packets [2].

In this chapter, we propose an efficient mutual authentication scheme for multihop-enabled PMIP networks, EM^3A [3], which thwarts authentication attacks, including DoS, colluding, impersonating, replay, and man-in-the-middle (MITM) attacks. We also offer a case study of applying our proposed scheme in a novel Multihop Authenticated Proxy Mobile IP (MA-PMIP) Scheme for Asymmetric VANET [4]. EM^3A thwarts authentication attacks when handovers occur through the Infrastructure-to-Vehicle-to-Vehicle (I2V2V) communications, while achieving reduced overhead. In addition, we present a key establishment scheme based on symmetric polynomials [5–7], which generates a shared secret key between MN and RN. Compared to existing authentication schemes, our proposed scheme achieves higher secrecy as well as lower computation and communication overheads. For a domain with n MAGs, our scheme achieves $t \times 2^n$-secrecy, whereas existing symmetric polynomial-based authentication schemes achieve only t-secrecy. Extensive simulation is performed

S. Taha and X. Shen, *Secure IP Mobility Management for VANET*,
SpringerBriefs in Computer Science, DOI: 10.1007/978-3-319-01351-0_3,
© The Author(s) 2013

Fig. 3.1 Infrastructure-
connected multihop mobile
network. MN_a is roaming
to a relayed communication
through relay node MN_b

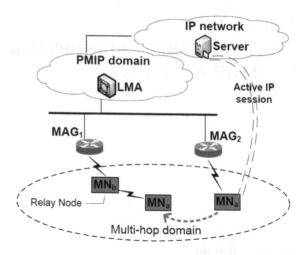

to show that our scheme can be used with seamless handover, since it results in low
authentication delay. In addition, the proposed key establishment scheme achieves
lower revocation overhead than that achieved by existing symmetric polynomial-
based schemes.

3.2 Related Work

For the problem of security and authentication schemes for multihop wireless networks [8], previous works have mainly focused on two different approaches: (1) end-
to-end authentication; and (2) hop-by-hop authentication.

3.2.1 End-to-End Authentication

The end-to-end authentication schemes employ a relay node (RN) to only forward
the authentication credentials between mobile node and the infrastructure. In [9], the
MN uses a trusted delegation entity and its public key certificate to authenticate itself
to the foreign gateway. On the other hand, the scheme in [10] uses both a symmetric
key for authenticating an MN to its home network, and a public key for mutual
authentication between home network and foreign network. However, the expensive
computation involved with public key operations tends to increase the end-to-end
delay.

Conversely, a symmetric key-based authentication scheme for multihop Mobile
IP is proposed in [11]. In that work, an MN authenticates itself to its home authentication server (HAAA) using the extensible authentication protocol [12]. After a

successful authentication, the HAAA derives a group of keys to be used by the MN, including a shared master key, extensible master key, and foreign MIP key. Despite the low computation and communication overheads, the symmetric key-based schemes cannot achieve as strong levels of authentication as those achieved by public key-based schemes. This is because the sharing of the secret key between the two peers increases the chances for adversaries to identify the shared key. Instead, public key-based schemes create a unique secret key for each user; hence, it is more difficult for adversaries to identify the keys.

3.2.2 Hop-by-Hop Authentication

The hop-by-hop authentication schemes implement authentication algorithms between each pair of hops in the routing path from the MN to the destination. A mutual authentication that depends on both secret splitting and self-certified schemes is proposed in [13]. However, both schemes are prone to DoS attacks. Another scheme for hop-by-hop authentication, called Alpha, is presented in [14]. In Alpha, the MN signs the messages using a hash chain element as the key for signing, and then delays the key disclosure until receiving an acknowledgement from the intermediate node. Although Alpha protects the network from insider attacks, it suffers from a high end-to-end delay. A hybrid approach, the adaptive message authentication scheme (AMA), is proposed in [15]. AMA adapts the strength of the security checks depending on the security conditions of the network at the moment of packet forwarding. AMA works under the assumption that the entire network cannot be attacked at the same time. In some spots, the adversary attacks all the transmitted messages, while in other spots there is no attack at all. Consequently, AMA proposes two different modes: a relaxed mode, to be used as default mode, and a check-all mode, which is used when attacking is discovered.

3.3 System Model

Different from the aforementioned authentication schemes, in this chapter we propose a light-weight mutual authentication scheme, EM^3A, to be employed between the mobile node and the relay, which mitigates the high delay that is introduced by previous hop-by-hop schemes. This then means that the proposed scheme can be used with seamless handover operations in multihop VANET during the I2V2V communications, as illustrated in Sect. 3.7.

3.3.1 Network and Communication Model

Consider an infrastructure-connected multihop mobile network such as that depicted in Fig. 3.1. The IP mobility support in MNs is provided by means of an adapted version of PMIP for multihop domains [16]. The only modification we introduce to [16] is the strict requirement for the MN to first connect directly to a MAG in order to obtain a valid IP prefix in the domain. Once an MN joins the domain for the first time, it sends Router Solicitation (RS) messages, which are employed by the MAG as a hint for detecting the new connection. After the PMIP signalling has been completed, the MAG announces the IP prefix in a unicast RA message delivered to the MN over the one-hop connection. After that, the MN may eventually divert to use an RN to reach the fixed network. We also assume that, after authenticating them, legitimate nodes in the PMIP domain faithfully follow the routing protocol when they are selected to provide their relay services for another MN in their surroundings.

The multihop communications that are studied in our system model are those occurring between MN and RN, when the MN intends to maintain a connection to the infrastructure. Applications of multihop mobile networks have been largely studied not only for vehicular communications networks, but also for wireless personal area networks [17], and wireless local area networks [18].

3.3.2 Threat and Trust Models

We consider both internal and external adversaries. Internal adversaries are legitimate users who exploit their legitimacy to harm other users. Thus, having the same capabilities as the legitimate users, internal adversaries have authorized credentials that can be used in the PMIP domain. Two types of internal adversaries are defined: impersonation and colluder. The former impersonates another MN's identity and sends neighbor discovery messages, such as Router Solicitation, through the relay node. The latter colludes with other domain users using their authorized credentials in order to identify the shared secret key between two legitimate users.

In the case of external adversaries, these are unauthorized users who aim at identifying the secret key and breaking the authentication scheme. These adversaries have high-quality monitoring devices so they can eavesdrop on messages transmitted between an MN and an RN. Moreover, they can inject their own messages and delete other authorized user's transmitted messages as well. We consider replay, MITM, and DoS attacks as external adversaries. The goal of the MITM and replay attacks is to identify a shared key between two legitimate users, while the goal of the DoS attack is to exhaust the system resources following a kind of irrational attack. DoS attackers can also be considered as internal adversaries when the attacker is one of the legitimate nodes.

In our model, we consider the LMA and all MAGs in the domain to be trusted entities. An MN trusts its first attached MAG in such a way that this MAG does not reveal the MN's evaluated domain polynomial, which is used by the MN to create shared keys with relay nodes. In addition, the MN trusts the LMA that maintains the secret domain polynomial, which can be used to reveal the shared keys for all nodes in the network.

3.3.3 Symmetric Polynomials

A symmetric polynomial is defined as any polynomial of two or more variables that achieves the interchangeability property, i.e., $f(x, y) = f(y, x)$. Such a type of mathematical function is often used by key establishment schemes to generate a shared secret key between two entities. A polynomial distributor, such as the access router, securely generates a symmetric polynomial and evaluates this polynomial with each of its users' identities. For example, given two users identities 1 and 2, and the symmetric polynomial $f(x, y) = x^2 y^2 + xy + 10$, the resultant evaluation functions are $f(1, y) = y^2 + y + 10$ and $f(2, y) = 4y^2 + 2y + 10$, respectively. Then, the polynomial distributor keeps the original polynomial secured, and sends the evaluated polynomials to each user in a secure way. Afterwards, the two users can share a secret key between them by calculating the evaluation function for each other. Continuing with the previous example, if user 1 evaluates its function $f(1, y)$ for user 2, it obtains $f(1, 2) = 16$. In the same way, if user 2 evaluates the function $f(2, y)$ for user 1, it obtains $f(2, 1) = 16$. Therefore, both users share a secret key, 16, without transmitting any additional messages to each other.

New decentralized key generation schemes are proposed in [6, 7] to generate a shared secret key between two arbitrary MNs that are located in two heterogeneous networks. These schemes achieve t-secrecy level, where t represents the degree of the generated polynomial. A scheme with t-secrecy property can be broken if $t + 1$ users collude to reveal the secret polynomial. Moreover, for only one MN's revocation, the decentralized schemes require changing the entire system's keys, which leads to a high communication overhead. Later in Sect. 3.6, we show how EM^3A reduces the revocation overhead, and increases the achieved secrecy level.

3.4 Efficient Mutual Multihop Mobile Authentication Scheme

Efficient mutual multihop mobile authentication scheme (EM^3A) consists of three main phases: a key establishment phase, for establishing and distributing keys; a mobile node registration phase, for MN's first attachment to the PMIP domain; and an authentication phase, for mutually authenticating the MN and RN.

3.4.1 Key Establishment Phase

Considering a unique identity for each MAG, the LMA maintains a list of those identities and distributes them to all legitimate users in the PMIP domain. The MAGs list's size depends on the number of MAGs in the domain. For n MAGs, each legitimate MN requires $(n \times \log n)$ bits to store this list. We argue that such storage space can be adequately found in mobile networks, such as vehicular networks. The LMA is also authorized to replace the identity of any MAG with another unique identity (this is specially useful for the management of MN's revocation, as it will be illustrated in Sect. 3.4.4).

Each MAG in the domain generates a four-variable symmetric polynomial $f(w, x, y, z)$, which we call the network polynomial, and then sends this polynomial to the LMA in its domain. After collecting all network polynomials, $f_i(w, x, y, z), i = 1, 2, ...n$, from every MAGs $(MAG_1, MAG_2, ..., MAG_n)$, the LMA computes the domain polynomial, $F(w, x, y, z)$, as follows:

$$F(w, x, y, z) = \sum_{i \in R^n}^{l} f_i(w, x, y, z), 2 \leq l \leq n \qquad (3.1)$$

where n is the number of MAGs in the domain. The LMA randomly chooses and sums l network polynomials from the received n polynomials in order to construct the domain polynomial. The reason for not summing all the network polynomials is twofold: increasing the secrecy of the scheme from t-secrecy to $t \times 2^n$-secrecy, and decreasing the revocation overhead at the time of MN's revocation. After constructing the domain polynomial $F(w, x, y, z)$, the LMA evaluates it for each MAGs identity, ID_{MAG}, individually. The LMA then securely sends to each MAG its corresponding evaluated polynomial. Later on, the evaluated polynomials, $F(ID_{MAGi}, x, y, z)$, with $i = 1, 2, ..., n$, are used to generate shared secret keys among arbitrary nodes in the domain.

3.4.2 MN Registration Phase

When an MN first joins the PMIP domain, it authenticates itself to the MAG to which it is directly connected. This initial authentication may be done using any existing authentication schemes, such as RSA. After guaranteeing the MN's credentials, the first-attached MAG securely replies by evaluating its domain polynomial, $F(ID_{MAG}, x, y, z)$, using the MN's identity, to obtain $F(ID_{MAG}, ID_{MN}, y, z)$. Afterwards, the LMA also sends the list of current MAGs's identities to the MN.

The MN stores the received list along with the identity of its first-attached MAG (ID_{FMAG}). As a result, a mobile node a can establish a shared secret key with another mobile node b in the same PMIP domain, by evaluating its received polynomial, $F(ID_{FMAGa}, ID_a, y, z)$, to obtain $F(ID_{FMAGa}, ID_a, ID_{FMAGb}, ID_b)$. Similarly, b evaluates its received polynomial, $F(ID_{FMAGb}, ID_b, y, z)$, to obtain $F(ID_{FMAGb}, ID_b, ID_{FMAGa}, ID_a)$. Since the domain polynomial, F, is a symmetric polynomial, the two evaluated polynomials result in the same value and they represent the shared secret key between mobile nodes a and b, K_{a-b}.

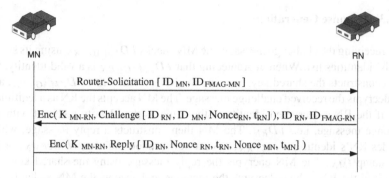

Fig. 3.2 EM^3A authentication phase

3.4.3 Authentication Phase

Figure 3.2 illustrates the MN-RN authentication phase. When an MN roams to a relayed connection, the neighbor discovery messages for movement detection in the multihop-enabled PMIP scheme must go through an RN. The goal of the authentication phase is to support mutual authentication between the roaming MN and the RN. After a successful authentication phase, the RN ensures that the MN is a legitimate user, and the MN ensures that the RN is a legitimate relay. The authentication phase is composed of the three stages described as below.

3.4.3.1 MN Initialization

The MN sends a Router Solicitation (RS) or Neighbor Solicitation message that includes its identity and its first attached MAG's identity, $ID_{FMAG-MN}$. Therefore, the intended RN checks its stored MAGs list to see if $ID_{FMAG-MN}$ is currently a valid identity. If there is no identity equals to $ID_{FMAG-MN}$, the RN rejects the MN and assumes it is a revoked or malicious node. Otherwise, if $ID_{FMAG-MN}$ is a valid identity, the RN continues with the next step to check the MN's authenticity.

3.4.3.2 Challenge Generation

By using the MN's identity and $ID_{FMAG-MN}$, the RN generates the shared key K_{MN-RN} as described in the registration phase. The RN then constructs a challenge message, which includes its own identity, ID_{RN}, the MN's identity, a random number $Nonce_{RN}$, and a time stamp t_{RN}. Finally, the RN encrypts the challenge message using the shared key, K_{MN-RN}, and sends it, along with ID_{RN} and its first attached MAG's identity, $ID_{FMAG-RN}$, to the MN.

3.4.3.3 Response Generation

After receiving the challenge message, the MN checks $ID_{FMAG-RN}$ using its stored MAGs' identities list. When guaranteeing that $ID_{FMAG-RN}$ is a valid identity, the MN reconstructs the shared key, by using the RN's identity and $ID_{FMAG-RN}$, and then decrypts the received challenge message. The MN accepts the RN as a legitimate relay if the RN's decrypted identity is the same as the identity received with the challenge message, i.e., ID_{RN}. The MN then constructs a reply message, which includes RN's identity, $Nonce_{RN}$, t_{RN}, a new random number $Nonce_{MN}$, and a time stamp t_{MN}. The MN encrypts the reply message using the shared key, and sends it to the RN, which decrypts the message and accepts the MN as legitimate user if the decrypted $Nonce_{RN}$ equals to the original random number that the RN sent in the challenge message.

Once the authentication phase is completed, the neighbor discovery messages are properly forwarded toward the MAG, which allows for the multihop-enabled PMIP to continue its operation as described in [16] and maintain seamless communications. In Fig. 3.2, Enc(K, M) represents an encryption operation of a message M using a key K. In addition, the Router-Solicitation, Challenge, and Reply are the three messages transmitted between the MN and the RN.

3.4.4 Mobile Node Revocation

To achieve backward secrecy, EM^3A should guarantee that a revoked MN does not use any of its previous shared keys to deceive the RN. When an MN is revoked, the LMA replaces this MN's first-attached MAG's identity, $ID_{FMAG-MN}$, with another unique identity, ID_{NFMAG}, and sends the new identity to all legitimate nodes in the domain. Subsequently, each legitimate node updates its stored MAGs list by replacing the old identity with the new one. The LMA also sends a message to each MAG in the domain, which includes a list of the mobile nodes that have ID_{NFMAG} as their first-attached MAG's identity, along with an evaluated polynomial, $F(ID_{NFMAG}, x, y, z)$, for the FMAG's new identity. Afterwards, the MAGs send the evaluated polynomial for those MNs that are in the received list and under MAGs' coverage areas. Eventually, each mobile node, in the MNs list, receives a new evaluated polynomial, $F(ID_{NMAG}, ID_{MN}, y, z)$, for both its identity and the new first-attached MAG's identity. Therefore, instead of changing the entire domain keys, only the MNs that share the same $ID_{FMAG-MN}$ need to change their evaluated polynomials and keys.

Figure 3.3 shows an example of the revocation operation. Consider a revoked MN_4 with a first-attached MAG as MAG_3. Three main messages are transmitted. The LMA first changes the MAG_3 identity to MAG_{03} and then broadcasts the new identity to all MNs in the domain. The second message transmitted from the LMA to all MAGs and it includes the new evaluated polynomial with the new identity, $F(ID_{MAG03}, x, y, z)$, along with the MNs that share the MAG_3 as their

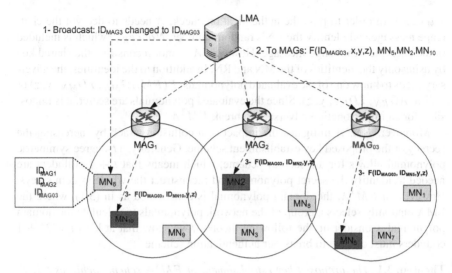

Fig. 3.3 EM^3A MN revocation

first-attached MAG. In the figure, MN_2, MN_5, and MN_{10} are those intended nodes. Note that the nodes sharing the first-attached MAG may not be located under the same MAG's control, therefore the LMA sends the second message to all MAGs. Finally, the last message is transmitted from the MAGs to those intended nodes, MN_2, MN_5, and MN_{10}.

3.5 Security Analysis

The security of our proposed scheme is based on the secrecy level of the key establishment phase proposed in Sect. 3.4.1. Therefore, in the following subsections we compute the security level of EM^3A scheme, and show that it thwarts both the internal and external adversaries defined in Sect. 3.3.2.

3.5.1 Internal Adversaries

The proposed EM^3A authentication scheme thwarts impersonation attacks by using a shared secret key, which is only known by the two communicating entities. To illustrate this, consider an adversary A, which aims at impersonating an MN in order to join a new MAG through an RN, and illegally benefit from the domain services. First, A sends an RS message and attaches the MN's identity, ID_{MN}. The RN replies with a challenge message, which is encrypted by the shared key

K_{MN-RN}. In order to pass the authentication check, A needs to decrypt the challenge message and identify the RN's random number, $Nonce_{RN}$, which is included in the encrypted challenge message. However, A cannot reconstruct the shared key by using only the identities of the MN and RN. In addition to the identities, the adversary needs to know one of the evaluated polynomials, $F(FMAG_{MN}, ID_{MN}, y, z)$ or $F(FMAG_{RN}, ID_{RN}, y, z)$. Since the evaluated polynomials are secret, it is impossible for an impersonation adversary to break EM^3A.

Moreover, EM^3A mitigates the impact of a collusion attack by increasing the secrecy of the proposed key establishment scheme. Generally, a t-degree symmetric polynomial allows for a t-secrecy scheme, which means that $t + 1$ colluders are needed to identify the secret polynomial and reconstruct the whole system's keys. However, in EM^3A, the domain polynomial is constructed as in (3.1), where the LMA randomly selects a group of the network polynomials to calculate the domain polynomial. Considering the following theorem, we show that at least $t \times 2^n + 1$ colluders must collude to break our authentication scheme.

Theorem 3.1 *The proposed key establishment in EM^3A scheme achieves $t \times 2^n$ secrecy level.*

Proof If we consider the secrecy of each network polynomial as t, then the secrecy s of the domain polynomial can be computed as follows:

$$\begin{aligned} s &= \sum_{k=2}^{n} \binom{n}{k} \times t \\ &= t \times \sum_{k=0}^{n} \binom{n}{k} - [\binom{n}{0} + \binom{n}{1}] \\ &= t \times [2^n - (1 + n)] \\ &\simeq t \times 2^n \end{aligned} \tag{3.2}$$

where n is the number of MAGs in the domain and t is the degree of network polynomials. Since the secrecy increases from t to $t \times 2^n$, the number of colluders that can break the scheme also increases from $t + 1$ to $(t \times 2^n) + 1$.

To show the significance of increasing the secrecy level, consider a PMIP domain with 10 MAGs and a symmetric polynomial of degree 10. On one hand, in traditional symmetric polynomial based key establishment schemes, only 11 colluders ($t + 1$) can break the system by revealing the used secret polynomial. On the other hand, in our scheme, EM^3A, the number of colluders increases to be 1001 $((t \times 2^n) + 1)$, which is 10-doubles of the original colluders. Consequently, as a way to impede the colluder attacks in our scheme, t is chosen to be a large number, and n should be preferably large.

3.5.2 External Adversaries

Similar to impersonation attacks, DoS attackers may trigger forged RS messages in order to exhaust the RN and MAG resources. Without EM^3A, the RN forwards all RS messages to the MAG and facilitates the DoS attack. However, using EM^3A, a DoS adversary A should know a valid shared key, K_{MN_i-RN}, in order for the RN to forward the RS message. Since A is an external adversary, it cannot construct any key, even if it knows the identity of a legitimate MN. On the other hand, A may repeat one of the RS messages that have been previously transmitted by a legitimated user, in order to trigger a replay attack. However, EM^3A thwarts this attack by adding both time-stamps and random nonces for each transmitted message between the MN and the RN. Finally, A may trigger an MITM attack in order to impersonate an MN or an RN. However, given that both the challenge and reply messages are encrypted, A cannot replace the MN or RN identities. Once more, A would need to know the shared key first in order to perform such attack.

3.6 Performance Evaluation

3.6.1 Computation and Communication Overheads

In this section, we evaluate the EM^3A scheme compared to previous multihop authentication schemes. Tables 3.1 and 3.2 show the computation and communication overheads for EM^3A comparisons. T represents the required time for an operation and B represents the transmitted bytes. Our scheme has the smallest computation overhead among other schemes, because EM^3A requires only two symmetric-key encryption operations ($2 \times T_c$). Both the AMA [15] and GMSP [10] require time for signing and verifying signatures (T_s, T_v), hence their computation overheads are higher than that in EM^3A. Like our proposed scheme, the multihop MIP scheme [11] consumes little time in computation; however, it requires high communication overhead to exchange a large number of keys. Moreover, ALPHA [14] requires an extra time ($T_{disclose}$) to delay the disclosure of the secret key. We employ Crypto++ benchmark [19] to compare the cost of each scheme. We use AES and RSA 1024 symmetric and public key operations respectively, in order to calculate the computation time required by the different schemes. The Round Trip Time (RTT) considered between vehicle and relay node is 5 ms.

Considering the communication overhead perspective, we observe that AMA, GMSP, and multihop MIP require transmission of a sender certificate in each transmitted message. Instead, the EM^3A scheme exchanges the list of MAGs only once at the key establishment phase, and the challenge/response messages ($B_{CHL-RESP}$) during handovers. The average length of the X.509 certificate is 3500 bytes, while the list of MAGs has a length of $n \log_2 n$ bits, where n is the number of MAGs in the PMIP domain. Therefore, in order for EM^3A to have a higher communication

Table 3.1 EM^3A computation overhead

Scheme	Computation overhead	Time (ms)
AMA [15]	$T_s + T_v \times Pr_{check}$	2.55
GMSP [10]	$T_s + T_v + T_c$	2.60
Multihop MIP [11]	$T_c + T_{EAP}$	0.0194
ALPHA [14]	$T_c + T_{disclose}$	7.5094
EM^3A	$2 \times T_c$	0.0194

Table 3.2 EM^3A communication overhead

Scheme	Communication overhead
AMA [15]	B_{cert}
GMSP [10]	B_{cert}
Multihop MIP [11]	$B_{EAP} + B_{key-exchange}$
ALPHA [14]	$B_{ACK} + B_{disclose}$
EM^3A	$B_{FMAGs-list}$

overhead than that in the other schemes, it would have to satisfy the condition $n \log_2 n \geq 28000$ bits $\times m$, where m is the number of transmitted messages in the certificate-based schemes. Consequently, n should be at least $236.64 \sqrt{m}$ to satisfy such a condition. However, since n is a fixed value, and m increases over time with the length of active sessions, n becomes much smaller than m with time. Therefore, the condition cannot be satisfied and EM^3A's communication overhead is clearly lower when compared with the certificate-based schemes. Even when a lightweight certificate is used instead of the X.509 certificate, the EM^3A communication overhead is still lower than those with certificate-based schemes where certificates need to be appended to each message transmitted to different relay in the network. Note that ALPHA [14] results in the smallest communication overhead, but it suffers from a $T_{disclose}$ delay in the computation overhead, which is required before disclosing the secret key.

3.6.2 Simulation Results

We evaluate the impact of EM^3A in the overall performance of the network when an MN experiences handovers that involve the use of RNs. Experiments were conducted through simulations using OMNET++ tool. The RTT between LMA and MAGs is fixed to 10 ms. A server for the downloading of data traffic is located in an external network, so that RTT between server and LMA is 20 ms. The MN is moving at different speeds that cause the frequency of handovers to vary from one every 10 s to one every 50 s (i.e., highly dynamic and slow changing scenarios). We consider the

Table 3.3 PMIPv6 network simulation parameters

PHY layer	2.4 GHz, 5.5 Mbps, 100 mW Tx power, −110 dBm sensitivity
MAC layer	802.11 ad hoc mode, 150 m radio range
Traffic type/rates	UDP/VBR video (mean 600 Kbps),
	VBR audio (mean 320 Kbps), CBR best effort 100 Kbps
Session time	∼3 min

worst-case scenario in which every time the MN handovers to a new MAG, it first connects to an RN, so that EM^3A authentication is required before the exchange of neighbor discovery packets and PMIP signalling may happen. Other details of the simulation parameters are provided in Table 3.3.

Figure 3.4a shows the average throughput obtained for the multihop-enabled PMIP, when the EM^3A scheme is de-activated and activated respectively. It can be observed that the authentication scheme does not impact communications negatively, and that the achieved performance is almost equivalent to that achieved when no authentication has been activated. Thanks to the registration phase, which is executed when every node first joins the PMIP domain. At the moment of handover, EM^3A requires only one RTT between MN and RN before allowing for the continuation of normal handover signalling (i.e., the forwarding of RS from MN to MAG, the PMIP signalling between MAG and LMA, and the router advertisement message sent back to MN). The downside of such registration phase is the overhead and storage required for sending and maintaining the list of current identities for all the MAGs in the domain.

To better illustrate the impact of EM^3A, we provide the details for the handover delay obtained during highly dynamic and slowly-changing scenarios in Fig. 3.4b. When the EM^3A has been activated the delay increases by ∼1.1 and ∼2.5 % in each scenario. Consequently, the low computation overhead of the symmetric key encryption/decryption operations makes the authentication process a light-weight mechanism for securely using multihop communications in PMIP domains.

Figure 3.5 illustrates the performance of the network in terms of packet losses for real time (audio and video) and best effort traffic. In general, the employment of the authentication scheme does not present a major impact compared to non-secure multihop PMIP. In the most demanding scenario, where handovers occur every 10 s, a low 0.03 % average increment among the three types of traffic results from the delay caused by the processing of EM^3A traffic. In the case of medium-to-slow changing scenarios, packet losses remain as low as 1 %, and EM^3A accounts only for a 0.01 % increment.

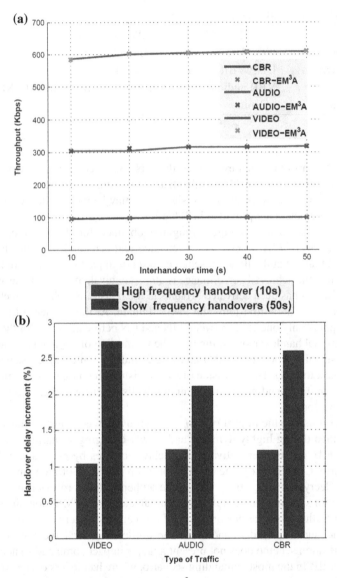

Fig. 3.4 Comparison of performance between EM^3A and non-secure multihop-enabled PMIP. **a** Average throughput. **b** Handover delay increase

3.7 Case Study: MA-PMIP for Asymmetric VANET

Multihop Authenticated PMIP (MA-PMIP) for Asymmetric VANETs [4] is a new proposed scheme in which we demonstrate that multihop paths are useful tools for improving the performance of infotainment applications and Internet access in

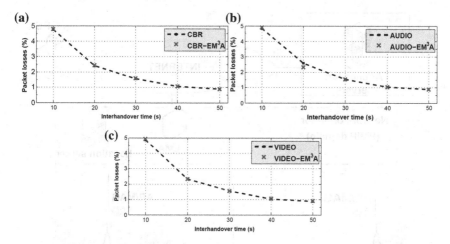

Fig. 3.5 Average packet losses obtained by EM^3A compared to non-secure multihop-enabled PMIP. **a** Packet losses for CBR traffic. **b** Packet losses for VBR audio traffic. **c** Packet losses for VBR video traffic

vehicular environments. Moreover, different from previous works that assume symmetric links among all wireless devices, we handle asymmetric links, and demonstrate that multihop communications are key for avoiding service breakage in the asymmetric VANET. In addition, and most importantly, we implement our efficient and mutual authentication scheme, EM^3A, and deal with the proposed MA-PMIP scheme to thwart authentication attacks when handovers occur through I2V2V communications.

The aforementioned contributions comprise the design goals of our Multihop Authenticated Proxy Mobile IP (MA-PMIP) scheme, which to the best of our knowledge, is the first to combine a predictive IP mobility scheme designed for multihop asymmetric VANET, with the security issues of employing I2V2V communications. In the following subsections, we first explain the network model for the MA-PMIP, then we briefly explain the handover operation through I2V2V communications in MA-PMIP, and finally we evaluate the computation overhead of our authentication scheme when implemented with MA-PMIP comparing to other previous schemes. For more detail about the novel MA-PMIP scheme, the reader is referred to [4].

3.7.1 MA-PMIP Network Model

We consider a vehicular communications network such as the one shown in Fig. 3.6. Connections to the infrastructure are enabled by means of road-side Access Routers (ARs), each one in charge of a different wireless access network. Vehicles are equipped with wireless interfaces, as well as GPS systems that feed a location service from which the location of vehicles is obtained. Beacon messages are employed

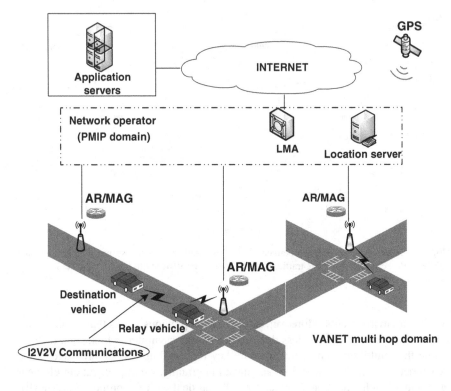

Fig. 3.6 MA-PMIP network model

by vehicles to inform about their location, direction, speed, acceleration, and traffic events to their neighbors.

The ARs employ a higher transmission power than the one employed by vehicles. Therefore, we consider the presence of asymmetric links in the VANET. The delivery of packets is assisted by a geographic routing protocol. To serve this protocol, a location server stores the location of vehicles, and is available for providing updated responses to queries made by the nodes' geo-networking layer. In order to forward packets within the multihop VANET, a virtual link between AR and vehicle is created [20]. That means that a geo-routing header is appended to each packet, where the location and geo-identifier of the recipient are indicated. In this way, the geo-routing layer is in charge of the hop-by-hop forwarding through multihop paths, with no need of processing the IP headers at the intermediate vehicles.

The ARs service areas are well-defined by the network operator. A well-defined area means that messages from ARs to the VANET are only forwarded within a certain geographic region [21]. Each AR announces its services in geocast beacon messages with the flag AccessRouter activated. The beacons are forwarded through multihop paths as long as the hops are located inside the coordinates indicated by the geocast packet header. In this way, vehicles in the connected VANET can extract information from the geocast header, such as AR's location, AR's geo-identifier, and the service area limiting coordinates. We assume the infrastructure is a planned network with non-overlapping and consecutive service areas. Note that, although service areas are consecutive, some locations within them are not reachable through one-hop connections. This may be caused by weak channel conditions, and by the asymmetric links between ARs and vehicles.

To ensure the proper operation of the geo-routing protocol and MA-PMIP, it is required to maintain state information at the entities exchanging IP packets. The following are the required data structures:

Neighbors table: stores information about the neighboring nodes. The table indicates a link type unidirectional or bidirectional for each neighbor. A node detects the bidirectional links in the following way: incoming links are verified when beacon messages are received from neighbors (i.e., this node can hear its neighbors); outward links are verified by checking the neighbors' locations and the node's transmission power, in order to calculate if such neighbors are inside the radio range (i.e., the neighbors can hear this node).

Default gateway table: stores information about the AR in the current service area. It contains the AR's geo-identifier and the service area coordinates. If the destination of a packet is an external node, the geographic routing forwards the packet toward the default gateway indicated in this table. Then, the AR routes the packet to its final destination.

We only consider IP-based applications accessed from the VANET. Such applications are hosted in external networks that may be private (for dedicated content), or public, such as the Internet. Since we have selected PMIP for handling the IP mobility in the network, all the ARs are assumed to belong to a single PMIP domain. The AR and MAG are co-located in our model. Therefore, the terms AR and MAG are used interchangeably in the following sections.

Unlike [21], in our scheme the AR does not send Router Advertisement (RA) messages announcing the IP prefix to vehicles in the service area. Instead, when a vehicle joins the network for the first time, individual IP prefixes are allocated through PMIP. It is required by MA-PMIP to obtain this initial IP configuration only when a one-hop connection exists between vehicle and MAG, so that authentication material is securely exchanged for future handovers of the vehicle over multihop paths. Note that a one-hop connection between two nodes is only established when a bidirectional link exists between them.

3.7.2 MA-PMIP Handover Operation

The signalling of MA-PMIP for initial IP configuration follows the standard PMIP. Once the vehicle joins the domain for the first time, it sends Router Solicitation (RS) messages, which are employed by the MAG as a hint for detecting the new connection. After the PMIP signalling has been completed, the MAG announces the IP prefix in a unicast RA message delivered to the vehicle over the one-hop connection.

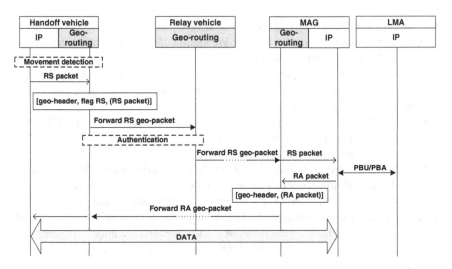

Fig. 3.7 MA-PMIP handover through I2V2V communications

Figure 3.7 shows the basic MA-PMIP signalling employed when a vehicle experiences a handover through a relay. The movement detection could be triggered by any of the following events: (1) the vehicle has started receiving AR geocast messages with a geo-identifier different from the one registered in the default gateway table; or (2) the vehicle has detected its current location falls outside the service area of the registered AR. If the vehicle loses one-hop connection toward the MAG, but is still inside the registered service area, then no IP mobility signalling is required and packets are forwarded by means of the geo-routing protocol.

After movement detection, the RS message is an indicator for others (i.e., relay vehicle and MAG) of the vehicle's intention to re-establish a connection in the PMIP domain. Thus, an authentication is required to ensure that both mobile router and relay are legitimate and are not performing any of the attacks described in Sect. 3.3.2. Therefore, we apply our proposed EM^3A scheme explained in Sect. 3.4 to implement the authentication required in MA-PMIP. Once the nodes are authenticated, the RS packet is forwarded until it reaches the MAG, and the PMIP signalling is completed in order to maintain the IP assignment at the vehicles new location. To take advantage of the location information in VANET, we propose a prediction mechanism that enables a timely handover procedure. It consists of an estimation of the time at which the vehicle will move to a new service area.

3.7.3 Authentication Evaluations

To measure and compare the impact of the MA-PMIP authentication mechanism, we have integrated an implementation of AMA [15], with a simplified version of a multihop PMIP scheme (i.e., MA-PMIP with our proposed authentication mechanism disabled).

Fig. 3.8 MA-PMIP
authentication delay

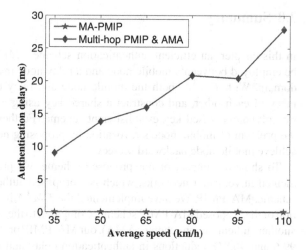

Fig. 3.9 MA-PMIP
communication overhead

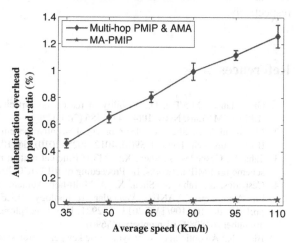

Figure 3.8 shows the authentication delay when the vehicle moves at different average speeds. Figure 3.9 depicts the comparison in terms of authentication overhead to payload ratio. As shown in both figures, MA-PMIP not only requires smaller delay and communication overhead than Multihop PMIP & AMA, but also has almost fixed impact for different speeds. On the other hand, Multihop PMIP & AMA have authentication delay and communication overheads that increase almost linearly with speed. Compared with Multihop PMIP & AMA, MA-PMIP achieves 99.6 and 96.8 % reductions in authentication delay and communication overhead, respectively. The reason for these reductions is the high computation and communication efficiency achieved by our proposed authentication scheme. Therefore, unlike Multihop PMIP & AMA, MA-PMIP can be used with seamless mobile applications, such as VoIP and video streaming.

3.8 Summary

In this chapter, an efficient authentication scheme, EM^3A, has been proposed to be employed between a mobile node and a relay node in a multihop-enabled PMIP domain. With EM^3A, both the mobile node and relay node guarantee the legitimacy of each other, and construct a shared key using a novel proposed symmetric polynomial-based key establishment scheme. Furthermore, we have mitigated the problem of mobile node's revocation by proposing new security steps that also achieve mobile node backward secrecy.

To show the impact of our proposed scheme, we present a case study implemented in vehicular networks, which is a proposed authentication multihop PMIP scheme, MA-PMIP. We have implemented the EM^3A in the multihop communication of the proposed MA-PMIP scheme. Compared to the AMA-PMIP that employed another authentication scheme (AMA), our MA-PMIP protocol with EM^3A achieves 99.6 and 96.8 % reductions in authentication delay and communication overhead, respectively.

References

1. Grossglauser, M., Tse, D.: Mobility increases the capacity of ad hoc wireless networks. IEEE/ACM Trans. Netw. **10**(4), 477–486 (2002)
2. Mahmoud, M.E., Shen, X.: PIS: A practical incentive system for multihop wireless networks. IEEE Trans. Veh. Technol. **59**(8), 4012–4025 (2010). doi:10.1109/TVT.2010.2062549
3. Taha, S., Céspedes, S., Shen, X.: EM3A: Efficient mutual multi-hop mobile authentication scheme for PMIP networks. In: Proceeding of IEEE ICC 2012. Ottawa, Canada (2012)
4. Céspedes, S., Taha, S., Shen, X.: A Multi-hop Authenticated Proxy Mobile IP Scheme for Asymmetric VANET, Vehicular Technology, IEEE Transactions on , vol. 99, pp.1–10. doi:10.1109/TVT.2013.2252931, http://ieeexplore.ieee.org/stamp/stamp.jsp?tp=& arnumber=6480892&isnumber=4356907
5. Blom, R.: An optimal class of symmetric key generation systems. In: Proceeding of the 84th Workshop on the Theory and Application of of Cryptographic Techniques, EUROCRYPT'85, pp. 335–338. Linz, Austria (1985)
6. Gupta, A., Mukherjee, A., Xie, B., Agrawal, D.P.: Decentralized key generation scheme for cellular-based heterogeneous wireless ad hoc networks. J. Parallel Distrib. Comput. **67**(9), 981–991 (2007)
7. Pillai, K., Sebastain, M.: A hierarchical and decentralized key establishment scheme for end-to-end security in heterogeneous networks. In: Proceeding of IEEE International Conference on Internet Multimedia Services Architecture and Applications, IMSAA 2009, pp. 1–6. Bangalore (2009). doi:10.1109/IMSAA.2009.5439493
8. Zhu, H., Lin, X., Lu, R., Ho, P., Shen, X.: Slab: a secure localized authentication and billing scheme for wireless mesh networks. IEEE Trans. Wireless Commun. **7**(10), 3858–3868 (2008)
9. Tang, C., Wu, D.: An efficient mobile authentication scheme for wireless networks. IEEE Trans. Wireless Commun. **7**(4), 1408–1416 (2008). doi:10.1109/TWC.2008.061080
10. Xie, B., Srinivasan, A., Agrawal, D.: GMSP: A generalized multi-hop security protocol for heterogeneous multi-hop wireless network. In: Proceeding of IEEE Wireless Communications and Networking Conference, WCNC 2006, vol. 2, pp. 634–639. Las Vegas, USA (2006). doi:10.1109/WCNC.2006.1683543

11. Al Shidhani, A., Leung, V.C.M.: Secure and efficient multi-hop mobile IP registration scheme for MANET-internet integrated architecture. In: Proceeding of IEEE Wireless Communications and Networking Conference, WCNC 2010, pp. 1–6. Sydney, Australia (2006). doi:10.1109/WCNC.2010.5506193

12. Catur Bhakti, M., Abdullah, A., Jung, L.: EAP-based authentication with EAP method selection mechanism. In: Proceeding of International Conference on Intelligent and Advanced Systems, ICIAS 2007, pp. 393–396. Kuala Lumpur, Malaysia (2007). doi:10.1109/ICIAS.2007.4658415

13. Jiang, Y., Lin, C., Shen, X., Shi, M.: Mutual authentication and key exchange protocols for roaming services in wireless mobile networks. IEEE Trans. Wireless Commun. 5(9), 2569–2577 (2006). doi:10.1109/TWC.2006.05063

14. Heer, T., Götz, S., Morchon, O.G., Wehrle, K.: Alpha: an adaptive and lightweight protocol for hop-by-hop authentication. In: Proceeding of The 4th ACM International Conference on Emerging Networking EXperiments and Technologies, ACM CoNEXT '08, pp. 23:1–23:12. Madrid, Spain (2008)

15. Ristanovic, N., Papadimitratos, P., Theodorakopoulos, G., Hubaux, J.P., Le Boudec, J.Y.: Adaptive message authentication for multi-hop networks. In: Proceeding of 8th International Conference on Wireless On-Demand Network Systems and Services, WONS 2011, pp. 96–103. Bardonecchia, Italy (2011). doi:10.1109/WONS.2011.5720206

16. Asefi, M., Cespedes, S., Shen, X., Mark, J.W.: A seamless quality-driven multi-hop data delivery scheme for video streaming in urban VANET scenarios. In: Proceeding of IEEE ICC 2011, pp. 1–5. Kyoto, Japan (2011). doi:10.1109/icc.2011.5962785

17. Lee, M.J., Zhang, R., Zheng, J., Ahn, G.S., Zhu, C., Park, T.R.: IEEE 802.15.5 WPAN Mesh standard-low rate part: meshing the wireless sensor networks. IEEE J. Sel. Areas Commun. 28(7), 973–983 (2010)

18. Hiertz, G., Zang, Y., Max, S., Junge, T., Weiss, E., Wolz, B., Denteneer, D., Berlemann, L., Mangold, S.: IEEE 802.11s: WLAN mesh standardization and high performance extensions. IEEE Netw. 22(3), 12–19 (2008). doi:10.1109/MNET.2008.4519960

19. Dai, W.: Crypto++ 5.6. 0 benchmarks. http://www.cryptopp.com/benchmarks.html

20. Choi, J., Khaled, Y., Tsukada, M., Ernst, T.: Ipv6 support for vanet with geographical routing. In: Proceeding of 8th International Conference on ITS Telecommunications, ITST 2008, pp. 222–227. Phuket, Thailand (2008). doi:10.1109/ITST.2008.4740261

21. Baldessari, R., Bernardos, C., Calderon, M.: Geosac-scalable address autoconfiguration for vanet using geographic networking concepts. In: Proceeding of 19th IEEE International Symposium on Personal, Indoor and Mobile Radio Communications, PIMRC 2008, pp. 1–7. IEEE, French Riviera, France (2008)

Chapter 4
Physical-Layer Location Privacy for Mobile Public Hotspots in a NEMO-Based VANET

4.1 Introduction

As an extension of MIPv6, NEMO protocol works appropriately for a scenario such as the one depicted in Fig. 4.1, where a Wi-Fi hotspot is deployed in public transportation (such as buses, trains, shuttles) and called a NEMO-based VANET [1–4]. In such networks, the OBU inside a vehicle also works as a Mobile Router (MR) to support a group of Mobile Network Nodes (MNNs), such as cell phones and PDAs, located inside the vehicle with required communications.

In this chapter, we modify the ideas of obfuscation and power variability to propose a strong physical-layer location privacy scheme, the fake point-cluster based scheme, that can be used in public hotspots for NEMO-based VANET. To the best of our knowledge, the fake point-cluster based scheme is the first to apply obfuscation, i.e., concealing, to a user's location by an exact location rather than a wide area. Unlike existing obfuscation schemes, which are employed in the current Location Based Service (LBS), our proposed scheme thwarts such a physical-layer attacker who tries to exploit the high-accuracy positioning schemes to define the sender's exact location. In addition, unlike current power variability schemes, our scheme changes the signal's power with respect to a specific reference point that we call a fake point, therefore, it is difficult to mitigate the impact of the power variabilities.

4.2 Preliminaries

4.2.1 Wireless Position Estimation

Our threat model relates to a physical-layer attacker who exploits positioning systems in order to reveal a sender's physical location from the received signal strength (RSS). Therefore, the wireless positioning systems [5] are illustrated in more detail,

S. Taha and X. Shen, *Secure IP Mobility Management for VANET*,
SpringerBriefs in Computer Science, DOI: 10.1007/978-3-319-01351-0_4,
© The Author(s) 2013

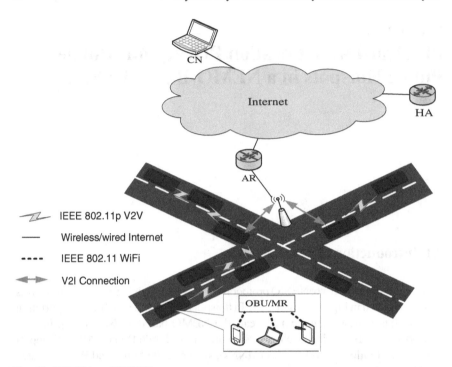

Fig. 4.1 NEMO-based VANET

in order to more deeply understand the attacker's strategy. In this sub-section, the two steps of the wireless position estimation process, distance measurement and location estimation, are described in detail.

The goal is to accurately estimate the mobile user's location inside a wireless network, such as Wi-Fi or a cellular network, when the user transmits signals. Starting with the distance measurement step, the mobile user's signal parameters are measured and the distances to the sender are estimated at certain reference points distributed across the network. Received signal strength (RSS), time of arrival (ToA), time difference of arrival (TDoA), and angle of arrival (AoA) are examples of the signal parameters. From the attacker's perspective, the RSS parameter is the best to use because unlike other signal parameter, RSS measuring requires only inexpensive equipment [5]. Therefore, in this sub-section, we focus on RSS-based estimation, in which each reference point at distance d from the mobile user measures the received signal power, $\bar{p}(d)$, as follows:

$$\bar{P}(d) = P_0 - 10n \log(d/d_0) \tag{4.1}$$

where P_0 is the received signal power to a known location that is located at distance d_0 from the reference point, and n is the path loss exponent, which depends on the propagation model of the signal in the wireless environment. In addition to the path-loss, the received power signal is also affected by both the shadowing and the fast fading (mutipath). In practice, with a long time interval of the signal observation, the effect of the multipath is excluded. Therefore, the received power is modeled to include the path-loss modeled in (4.1), and the shadowing modeled as a zero mean Gaussian random variable with a variance σ^2. The RSS measurement can be modeled as follows:

$$P(d) \sim N(\bar{P}(d), \sigma^2) \tag{4.2}$$

After measuring the RSS at a reference point i located in (x_i, y_i), the estimated distance, $f_i(x, y)$ to the sender is measured as follows:

$$f_i(x, y) = \sqrt{(x - x_i)^2 + (y - y_i)^2} \tag{4.3}$$

In the second step, location estimation, two techniques for location estimation are defined: (1) mapping (fingerprinting); and (2) geometric and statistical. The mapping techniques rely on an off-line training phase in which a database of different RSS estimations and their correspondent senders' locations is created. Depending on the training phase, a mapping method is used to match a new measured RSS value to entities in the database. In our NEMO-based hotspot, we assume that attackers cannot perform the training phase. Still alternatively, the geometric and statistical techniques can be used. In geometric techniques, the position of the mobile node (MN) can be estimated as the intersection of position circles obtained from RSS measurements that are estimated at different reference points. Since each RSS forms a circle, at least three reference points are needed to define the intersection point. This process is called triangulation, depicted in Fig. 4.2. In addition, using the statistical techniques, the location of the MN can be defined as follows:

$$Z = f(x, y) + \eta \tag{4.4}$$

where $Z = [Z_1, Z_2, \ldots, Z_N]^T$, $f(x, y) = [f_1(x, y), f_2(x, y), \ldots, f_N(x, y)]^T$, and $\eta = (\eta_1, \eta_2, \ldots, \eta_N)^T$ are the parameters collected from each reference point i as follows:

$$Z_i = f_i(x, y) + \eta_i, i = 1, 2, \ldots, N \tag{4.5}$$

where N is the number of reference points, and $f_i(x, y)$ is the distance that a reference point i estimates for the sender location (x, y) by using the measured RSS value as in (4.3), and η_i is the estimation error at this reference point.

After collecting the estimated distances from all reference points, a general esti-mation $\theta = [x, y]^T$ of an MN's location is calculated. Based on knowledge of the probability density function (pdf) of the estimation error, η, parametric or non-parametric techniques can be used. Non-parametric techniques such as fingerprinting are employed if the error's pdf is not defined, while parametric techniques such as Bayesian and maximum likelihood (ML) estimators are used when the error's pdf is known. The Bayesian approach is used in the presence of a prior pdf on θ, $\pi(\theta)$ in

Fig. 4.2 Attack triangulation
on MNN

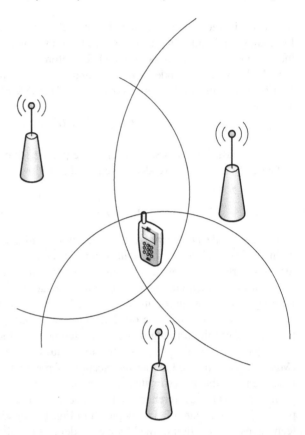

order to minimize the cost function of estimating θ by using either the minimum mean
square error, $\hat{\theta}_{MMSE}$, or the maximizing posterior estimations, $\hat{\theta}_{MAP}$ as follows:

$$\hat{\theta}_{MMSE} = E\{\theta \mid Z\} \tag{4.6}$$

$$\hat{\theta}_{MAP} = \arg\max_{\theta} P(Z \mid \theta)\pi(\theta) \tag{4.7}$$

On the other hand, the ML estimation is used when $\pi(\theta)$ is unknown, to maximize
the likelihood function.

$$\hat{\theta}_{ML} = \arg\max_{\theta} P(Z \mid \theta) \tag{4.8}$$

$$P(Z \mid \theta) = P_{\eta}(Z - f(x, y) \mid \theta) \tag{4.9}$$

where P_{η} is the conditional pdf of an estimation error condition on θ.

In RSS-based estimation, the error vector is assumed to be independent and is modeled as a zero-mean Gaussian. Therefore, (4.9) can be written as follows:

$$P(Z \mid \theta) = \prod_{i=1}^{N} P_{\eta_i}(Z_i - f_i(x, y) \mid \theta) \tag{4.10}$$

$$P_{\eta_i} = \frac{1}{\sqrt{2\pi}\sigma_i} \exp\left(-\frac{\eta_i^2}{2\sigma_i^2}\right) \tag{4.11}$$

$$P(Z \mid \theta) = \frac{1}{(2\pi)^{N/2} \prod_{i=1}^{N} \sigma_i} \exp\left(-\sum_{i=1}^{N} \frac{\eta_i^2}{2\sigma_i^2}\right) \tag{4.12}$$

Hence the ML estimator can be calculated as follows:

$$\hat{\theta}_{ML} = \arg \min_{[x,y]^T} \sum_{i=1}^{N} \frac{(Z_i - f_i(x, y))^2}{\sigma_i^2} \tag{4.13}$$

However, if we assume correlated Gaussian error components instead of independent components, the estimated ML can be written as follows:

$$\hat{\theta}_{ML} = \arg \min_{[x,y]^T} (Z - f(x, y))^T \Sigma^{-1} (Z - f(x, y)) \tag{4.14}$$

Figure 4.3 shows the flowchart of an RSS-based positioning system.

4.3 Related Work

4.3.1 Power Variability

Due to the open nature of wireless networks, hiding the transmitted wireless signals, and hence achieving physical layer location privacy is considered a challenging goal. In location privacy attacks, the attacker localizes the victim MNN by measuring its RSSs at certain reference points, as illustrated in Sect. 4.2.1. To thwart these attacks, [6] suggests employing a scheme in sensor networks called Hyberloc. In this scheme, the anchor nodes protect their location from the un-trusted nodes, while trusted nodes can easily localize those anchor nodes. The main idea of Hyberloc is to randomly choose a power for transmitting signals and attach this random power value in the transmitted encrypted packets. Therefore, having a shared key, only trusted nodes can identify the true sender's location. However, changing the transmission power values is considered to provide only weak location privacy, because the attacker

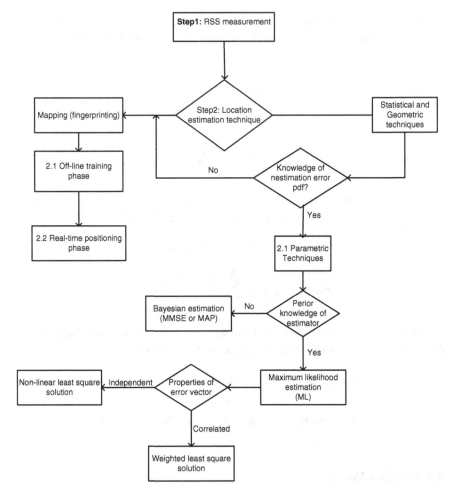

Fig. 4.3 Wireless position estimation

can easily fix these changes by multiplying the RSS at all monitor devices by a factor. In our proposed scheme, in addition to changing power levels as is done in Hyberloc, we confuse the attacker's monitoring devices by letting their measured RSSs be equalized for different MNNs; therefore, it becomes difficult for the attacker to mitigate the increase in power.

4.3.2 Noise Addition

Another scheme, hidden anchor, which relies on adding noise to the transmitted signals, is proposed in [7]. In this scheme, the anchor nodes use their neighbors' identities to hide their own identities from distrusted nodes, and at the same time encrypt and attach their real identities in the transmitted packets sent to trusted nodes. However, changing the nodes' identities does not achieve a sender's physical-layer privacy; rather, it helps in achieving link-layer location privacy. In addition, both anchor and trusted nodes add noise to their transmitted messages in order to prevent un-trusted nodes from measuring the RSSs and revealing their locations. However, adding noise to the transmitted messages affects transmission quality.

4.3.3 Obfuscation

Obfuscation, i.e., concealment, proposed in [8], is another way to protect a user's location privacy from location-based servers (LBSs). The idea of the obfuscation is to replace the real location information with fake information in order to decrease the accuracy of the localization process employed by LBS, and hence increase a user's location privacy. Three obfuscation techniques are proposed in [8]: enlarged area, shifted center, and reduced radius. In location-based applications, the user's location returned to the LBS represents an area rather than a specific location, therefore, obfuscation schemes are used to hide the true information about that area. However, these obfuscation schemes are not appropriate for Wi-Fi scenarios where the adversary gets a specific MNN's location rather than an area. In our proposed scheme, we modify the idea of obfuscation in order to return a wrong location point rather than a wide area.

With the goal of achieving obfuscation for users' information, [9] achieves user identity, time, and location obfuscation. User identity obfuscation, concealing the identity, is carried out by frequently changing a user's pseudonymity, while time obfuscation, concealing transmission time, is carried out by applying a silent period to thwart pseudonym correlation attacks. The silent period is defined in such a way as to increase a user's privacy level and hence decrease the positioning system's accuracy. Unlike the identity and time obfuscation that are mainly employed for link-layer obfuscation, location obfuscation is employed to achieve physical-layer location privacy. Assuming a fingerprinting positioning system, [9] achieves location obfuscation by proposing a silent Transmit Power Control (TPC) scheme that reduces transmission power at each user. Therefore, the number of APs that detect the transmitted signals decreases, as does the accuracy of the attacker's localization. The challenge of silent TPC is to allow users to change their transmission power without exchanging any information with their APs. Our proposed cluster-based scheme employs the same idea of TPC to reduce a user's transmission power. However, unlike our proposed scheme, the silent TPC scheme considers location attackers located only in neighbor networks rather than those located in the user's current network.

4.3.4 Smart Antennas

In [10], two strategies for a user's location privacy have been proposed with a main idea of using a smart antenna that emits a directional radiation pattern instead of using isotropic antennas. In the first strategy, using a smart antenna, the MNN maximizes the transmission power of the signals directed to the AP located in its network while preventing other APs from receiving any signals transmitted from this MNN. Therefore, other APs cannot triangulate this MNN and hence fail to reveal its location. On the other hand, if an MNN fails to prevent at least four APs from receiving its signals, then the MNN tries the second strategy in which the MNN maximizes the RSS localization bias at the APs around this MNN. By increasing the localization bias, the MNN guarantees that its surrounding APs estimate its position wrongly. To achieve the first strategy, the MNN first listens to the periodically received beacon packets that are transmitted by the nearby APs. The MNN then passively measures the RSSs of these beacon packets to estimate the APs' locations. However, if the APs change their power levels, then the MNN cannot estimate their locations and hence fails to protect this MNN's location privacy. In addition, the assumption of having a smart antennas in all MNNs is not reasonable due to their high cost.

4.3.5 Silent Period

In [11], a scheme called silent period is used to achieve physical and link-layer location privacy. It thwarts correlation attacks, so an attacker cannot relate two pseudonyms to the same MNN. A silent period is defined as a constant period, followed by a variable length period, in which an MNN changes its pseudonym and then keeps silent, not sending any messages. When an MNN starts sending frames after the silent period, the attacker cannot correlate between the MNN's new and old pseudonyms. However, this scheme degrades network performance when the MNN stops its transmission for some periods. In addition, a precise duplicate address detection scheme should be employed to ensure that the new pseudonym does not conflict with any other addresses in the network.

4.3.6 Phantom

Phantom is another scheme proposed in [12], to achieve sender physical-layer location privacy by creating a group of ghost transmitters, which retransmit the original transmitter's messages. Therefore, the attacker that uses an RSS-based fingerprinting localization scheme to localize the original transmitter receives a combination of both original signals and ghost signals. The power of the phantom comes from the inability of the adversary to distinguish between those signals. Although phantom achieves a high level of privacy, it also adds a large overhead when the number of ghosts and hence the energy consumed increase.

4.4 System Models

4.4.1 Network Model

A NEMO-based public hotspot is installed inside a large van, which in turn, constructs VANET communications with its neighbor vehicles, as depicted in Fig. 4.1. In addition to running a VANET routing protocol, the OBU of this van also works as a NEMO Mobile Router (MR) and runs a NEMO BS protocol; hence, it is denoted as OBU/MR. Inside the large van, Mobile Network Nodes (MNNs) represent different mobile devices such as cell phones, PDAs, and laptops.

We employ a MANET-centric approach to integrate NEMO and VANET protocols; therefore, only OBU/MR implements a NEMO BS protocol in addition to the MANET routing. All neighbor vehicles that are located on the OBU/MR-RSU path, including the RSU, implement only a MANET routing scheme, such as georouting protocols, as illustrated in Fig. 1.5a.

The communications among an OBU/MR and MNNs are generally structured using the IEEE 802.11 standard to form a Wi-Fi network, while the communications among OBU/MR and the road-side access points, which are employed to support the OBU/MR with the Internet connectivity, are applied by applying the NEMO BS protocol in VANET. In this paper, we focus on the communications of a vehicle's Wi-Fi network, which indeed are affected by the NEMO-VANET communications outside the vehicle.

Due to the varieties of link-layer connections in NEMO-based VANET, as illustrated in Fig. 4.1, three different MR-passengers communications types can be found in the Wi-Fi hotspot: in-vehicle, neighbor vehicles, and nested communications. The in-vehicle communications, the focus of this paper, are constructed among the in-vehicle MR that works as a hotspot's AP and passengers' devices inside the same vehicle. Neighbor vehicle communications can be created among an OBU/MR inside one vehicle and some passengers's devices inside neighbor vehicles. This kind of communication relies on the connectivity between vehicles; however, due to the diversity of vehicles' speeds and mobility models, neighbor communications face connections intermittencies, which lead to a degradation in network performance. Nested communications, also called nested-NEMO, are formed among a vehicle's MR and some passengers' devices under the control of another MR, which in turn, is under the control of this vehicle's MR.

In our model of a hotspot, The OBU/MR is located in the front of the vehicle and controls the whole hotspot, while all other MNNs are located randomly in the van and the transmission power signal of OBU/MR is considered to be much higher than those of MNNs. In addition, considering the same transmission environment for all

MNNs, we assume Gaussian noise with zero mean and σ^2 variance for all signals propagated inside the hotspot.

In addition, we assume that the OBU/MR logically divides the hotspots into k grid points, and attaches them to its periodically transmitted beacon, therefore, MNNs inside the hotspot use those grid points to implement our proposed scheme, as illustrated in Sect. 4.5.

4.4.2 Threat and Trust Models

A passive physical layer location privacy attacker deploys monitoring devices inside the whole network, in order to detect any transmitted signal and estimate the location of the sender using the received signal strength as illustrated in Sect. 4.2.1. The attacker's monitoring devices are assumed to have high sensing and processing capabilities, and their positions in the network can be changed by the attacker. Using the measured RSSs, each monitoring device estimates and transmits the distance to the intended sender to the attacker. Employing an ML estimation technique, the attacker uses the received distance estimations from all monitoring devices to estimate the exact location of the MNN. For more information about the ML statistical technique, the reader is referred to Sect. 4.2.1.

To attach itself to a hotspot, each MNN authenticates itself to the hotspot's MR and shares a secret key in order to encrypt its data-link frames, including its MAC address. Being unable to decrypt the transmitted frames' MAC addresses, the attacker depends only on the RSS measurements to localize the MNN. Many data-link authentication and location privacy schemes [13–20] can be used to secure a data-link layer's frames.

To apply a mutual authentication scheme among the MNNs and the OBU/MR, the OBU/MR periodically transmits its public-key certificate inside the hotspot, and we consider that there exists at least one online certificate verification server whom MNNs trust and use to verify the OBU/MR's certificate. Due to the multihoming technology that enables mobile devices to simultaneously attach to different networks, the MNNs can access the online certificate verification by an alternative Internet connection other than the mobile hotspot connection. For example, a cell phone can use its cellular network to connect to the Internet and verify the received certificates.

4.5 Fake Point-Cluster Based Physical Layer Location Privacy Scheme

The proposed fake point-cluster based scheme is a combination of two sub-schemes, fake point and cluster based, that can be employed individually to provide physical-layer location privacy for MNNs inside a NEMO-based VANET hotspot. The fake point sub-scheme achieves a higher location privacy level if the attacker's monitoring devices are located at the selected fake points' locations, while the cluster based sub-scheme achieves a higher location privacy when preventing attacker's monitoring devices from detecting the transmitted signals. In Sect. 4.6, we show that the proposed fake point-cluster based scheme increases the MNN's location privacy level. In the next subsections, fake point and cluster-based sub-schemes are presented, and then a scheme for their combination is explained.

4.5.1 Fake Point Location Privacy Sub-Scheme

The proposed fake point location privacy scheme is employed to protect MNNs' physical location privacy from insider passive attacks, which are explained in Sect. 4.4.2. The main idea is that, inside the hotspot, the MNNs select random locations, called fake points, that are used to confuse the attacker. The MNNs consider these fake points when calculating their transmission signals power. Therefore, if an attacker's monitoring devices are located at these fake points, then the measured RSS values at the monitoring devices are similar for all MNNs selecting the same fake point. In Sect. 4.6.2, the probability of having at least two MNNs choose the same fake point's location that contains an attacker's monitoring device is calculated. Therefore, these monitoring devices encounter some error when estimating the distances to MNNs. Depending on the error, the overall MNNs' location estimations also have some deviations, and hence the MNNs' location privacy is ensured.

4.5.1.1 Bootstrapping the Hotspot

Working as an AP, the OBU/MR broadcasts inside its network some beacon frames that contain its location, $(X_{OBU/MR}, Y_{OBU/MR})$, and a unique received signal power, P_u, that all MNNs in the network must consider when calculating their transmission signal powers. Using the AP's location and the required received power, the MNNs can define the distances to their AP and hence calculate appropriate transmission signal powers. The beacon frames also contain the mobile network prefixes (MNPs) for each MNN to select a unique MNP, and hence to be able to attach to the MENO, AP's certificate (CERT) for the MNNs, to check the authenticity of the AP. An authentication scheme such as in [13] can be used to achieve mutual authentication between MNNs and AP. After successful mutual authentication between the MNN and AP, the AP virtually divides the hotspots into K spatial grid points and securely sends the grid points' list to the authenticated MNN. In the remainder of this chapter, we use AP, OBU/MR, and MR interchangeably to represent the Wi-Fi AP.

4.5.1.2 MNN Attachment

To connect to the available hotspot, the MNN first calculates the distance d_{MNN-AP} to its AP as follows:

$$d_{MNN-AP} = \sqrt{(X_{MNN} - X_{AP})^2 + (Y_{MNN} - Y_{AP})^2} \qquad (4.15)$$

where (X_{MNN}, Y_{MNN}) is the MNN's current location measured by the MNN's GPS. Using this calculated distance and the required received power at AP, P_u, the MNN calculates its transmission power, P_{tr}, as follows:

$$P_u = \alpha - 10\beta \log(d_{MNN-AP}) \qquad (4.16)$$

where β is the path loss and α is a function of the transmission power P_{tr}. Instead of using the calculated transmission power, P_{tr}, the MNN uses another power, $\acute{P}_{tr}$, calculated related to a fake location that the MNN selects in the next step.

4.5.1.3 Identifying Fake Point

Inside its network, the MNN randomly selects a location, which we will call the fake point, from the grid points list that the AP securely sends to the authenticated MNN as mentioned in the Bootstrapping phase. Therefore, the fake point is one of the K spatial grid points that is defined by the AP, and represents a location (x, y) inside the hotspot. Using (4.15), the MNN recalculates its distance to the AP as the sum of the MNN-fake point distance and the fake point-AP distance, and then the MNN employs this distance to recalculate the transmission power, $\acute{P}_{tr}$ using (4.16). Therefore, the MNN recalculates the transmission power, $\acute{P}_{tr}$, in such a way that the MNN's signal is transmitted first to this fake point then to the AP. However, in our scheme, the MNN does not send its signals to this fake point; rather, it sends the signals directly to its AP. This deceiving action is only to confuse attackers. As depicted in Fig. 4.4, the distance between the MNN and its AP, d_{MNN-AP}, is always less than the sum of the distances of MNN-fake point, d_{MNN-F}, and fake point-AP, d_{F-AP}. Consequently the power transmitted from the MNN to the AP directly, P_{tr}, is less than that goes through the fake point, $\acute{P}_{tr}$. The main goal of selecting a fake point in the network is to have a possibility that one of the attacker's monitoring devices is located at this fake point. Therefore, when many MNNs select the same fake point, and an attacker's monitoring device is located in this fake point as depicted in Fig. 4.5, the estimated distances calculated by this monitoring device encounter much more estimation error then those calculated by other monitoring devices. Since the measured RSSs at the monitoring device are functions of the distances to the MNN, the recorded error increases as the distance between MNNs that selected the same fake points increases. In Sect. 4.6, the error encountered at the monitoring devices is measured to show the strength of the MNN's location privacy when using our proposed scheme.

Fig. 4.4 Fake point selection

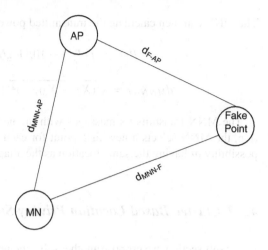

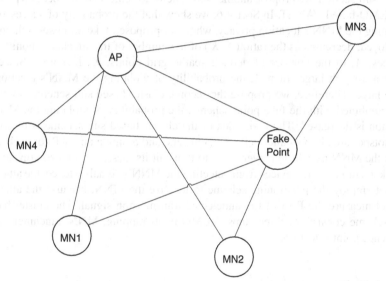

Fig. 4.5 MNNs select same fake point

To calculate the transmission signal power, $\acute{P}_{tr}$, the MNN randomly selects a fake point, (X_F, Y_F). Given the received signal power at AP is P_u, the signal power at the fake point, P_{ftr}, can be calculated as follows:

$$P_u = f(P_{ftr}) - 10\beta \log(d_{F-AP}) \qquad (4.17)$$

$$d_{F-AP} = \sqrt{(X_F - X_{AP})^2 + (Y_F - Y_{AP})^2} \qquad (4.18)$$

The MNN can then calculate the transmitted power to the fake point as follows:

$$P_{ftr} = f(\acute{P}_{tr}) - 10\beta \log(d_{MNN-F}) \tag{4.19}$$

$$d_{MNN-F} = \sqrt{(X_F - X_{MNN})^2 + (Y_F - Y_{MNN})^2} \tag{4.20}$$

The MNN transmits its messages with the new calculated power, $\acute{P}_{tr}$. In addition, the MNN selects a new fake point for each transmitted signal; therefore, the possibility of having the same location as the attacker's location will be increased.

4.5.2 Cluster-Based Location Privacy Sub-Scheme

In this sub-section, we propose another sub-scheme to achieve MNN location privacy in NEMO-based VANET. In Sect. 4.6, we show that the probability of successfully violating the MNN's location privacy, when our proposed fake point sub-scheme is employed, decreases as the ratio (A/K) of the number of the attacker's monitoring devices, A, to the number of defined spatial grid points, K, increases. Since K is always much larger than A, the probability of violating the MNN's location is quite large. Therefore, we propose the second, cluster-based sub-scheme, so when it is combined with the fake point scheme, the probability of violating the MNN's location is decreased. The main idea of the cluster-based sub-scheme is to divide the hotspot area into smaller cells, i.e., clusters, and assign a new AP for each cell. Thus the MNN uses little power value to transmit its messages, and so prevents an attacker's monitoring devices from detecting the MNN's signals. Hence, the attacker cannot employ the positioning scheme to localize the MNN, because the attacker cannot measure the RSS of the undetected transmission signal. The cluster-based sub-scheme consists of three steps: NEMO bootstrapping, MNN attachment, and reference point selection.

4.5.2.1 NEMO Bootstrapping

At the time of constructing the Wi-Fi as NEMO-based VANET communications, the OBU/MR that works as an AP for the whole network divides the network area into smaller n sub-areas called cells, $c_1, c_2, \ldots, c_n$. For each cell, c_i, the OBU/MR assigns an AP, which is a reference point, RP_i, that works as a local AP for all MNNs located within a distance r around RP_i. Considering each cell's coverage area as a circle, r represents the cell's radius, and we assume that all cells have the same radius, and they may overlap with each other as depicted in Fig. 4.6. From an attacker's perspective, we consider that there is at most one attacker's monitoring device in each cell.

Fig. 4.6 Cluster based sub-scheme

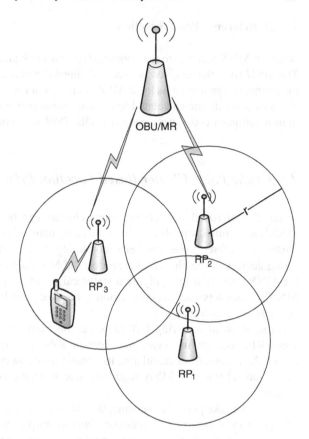

4.5.2.2 MNN Attachment

Working as a local AP, each RP broadcasts a beacon packet so only MNNs under its coverage area receive this beacon. The beacon message contains information about the RP, including its identity, ID_{RP}, its coverage area's radius, r, and its required received signal's power, P_{RRP}.

Considering the knowledge of its location, (X_{MNN}, Y_{MNN}), the MNN calculates the transmission signal power for its messages directed to its chosen RP, as follows:

$$FSPL(DB) = 20\log_{10}(r) + 20\log_{10}(f) + 32.45 \tag{4.21}$$

$$TP_{MNN} = P_{RRP} \times FSPL \tag{4.22}$$

where $FSPL$ is the free space path loss [21], which depends on the cell's radius in meters, r, and the transmitted signal frequency in megahertz, f.

4.5.2.3 Reference Point Selection

When an MNN attaches to the hotspot, it receives m beacons from m different RPs. The MNN sorts the received m beacons' signals' powers, and chooses the RP with the strongest signal to be its local AP. As depicted in Fig. 4.6, the MNN transmits all its messages with the calculated low transmission power to the selected RP, which in turn, retransmits the messages to the OBU/MR that works as the hotspot's AP.

4.5.3 Fake Point-Cluster Based Location Privacy Scheme

In our cluster based sub-scheme, since clusters can be spatially overlapped, the MNN's transmitted signals may be received by many clusters, not only the intended cluster. Therefore, if one attacker's monitoring device is in each cluster, the monitoring devices in the clusters that receive the MNN's signals can collude to reveal the MNN's location by applying a statistical positioning scheme. To increase the MNN's location privacy, a combination of the fake point based and the cluster based sub-schemes can be applied.

In addition to receiving OBU/MR beacon messages, the MNN also receives some RPs' beacon messages that contain RPs' positions, (X_{RP_i}, Y_{RP_i}), where $i \in \{1, 2, \ldots, m\}$. After calculating its transmission power as depicted in the cluster-based sub-scheme, the MNN randomly selects a fake point that is located in its cluster.

Using the fake point sub-scheme, the MNN calculates the required power at the fake point and then adjusts its transmit power to this power. Therefore, the MNN confuses some of the attacker's monitoring devices, and hence increases the estimation error resulting from the attacker's monitoring devices' collusion.

This combination between the fake point and the cluster-based sub-schemes prevents some attacker's monitoring devices located inside neighbor clusters from detecting the sender's transmitted signals. In addition, the fake point-cluster based selects a fake point inside the sender's cluster, in order to ensure higher location privacy and consume lower power.

4.6 Analytical Location Privacy Evaluation

In this section, an analytical analysis for the proposed scheme, the fake point-cluster based scheme, is presented. Similar to the evaluation analysis in [22], we employ three metrics: correctness, accuracy, and certainty. Correctness measures the additional estimation error that is added by our proposed scheme; accuracy measures the probability of an attacker's success in breaking the MNN's location privacy, and certainty measures the entropy of the achieved privacy.

4.6.1 Correctness

4.6.1.1 Fake-Point Based Sub-Scheme

When m different MNNs choose the same fake point, which in turn may be a location of an attacker's monitoring device, the attacker estimation error for the MNN's localization increases. Using the signal propagation model from [23], the MNN's RSS can be calculated as follows:

$$RSS = \frac{AP_t}{B + d^\alpha + C(\log_{10} d)^\beta + D} \tag{4.23}$$

where A, B, C, D, α, and β are a signal's parameters that can be estimated by the attacker. However, the attacker cannot estimate the MNN's transmission power, P_t, while the distance, d, to the target MNN is also unknown. Using Frii's formula, P_t can be expressed as:

$$P_t = RSS - G_t - G_r - 20\log_{10}\frac{\lambda}{4\pi d} \tag{4.24}$$

where G_t and G_r are the sending and receiving channel's gains. Therefore, substituting (4.24) in (4.23), RSS can be written as:

$$RSS = \frac{A(G_t + G_r + 20\log_{10}\frac{\lambda}{4\pi d})}{A - B - d^\alpha - C(\log_{10} d)^\beta - D} \tag{4.25}$$

By (4.25), the attacker can measure the RSS values, $RSS_1, RSS_2, \ldots, RSS_m$, for m different MNNs that select the same fake point in the fake point sub-scheme, as follows:

$$RSS_i = \frac{A(G_t + G_r + 20\log_{10}\frac{c}{4\pi f d})}{A - B - d^\alpha - C(\log_{10} d)^\beta - D} \tag{4.26}$$

where c is the speed of light and f is the signal's frequency, which is one of the channel parameters. Assuming 2.4 GHZ is the frequency band that is used in the hotspot, m MNNs select any of the 14 channels that are assigned in this band. The channels' frequency bands are spaced 5 MHZ apart; therefore, the differences in the term $\log_{10}\frac{c}{4\pi f d}$ for each MNN are negligible. Thus, (4.26) yields the same RSS's value for all MNNs that share the same fake point.

From (4.26), the attacker estimates the distance between its location, which is the fake point's location, (x_f, y_f), and the target-point's location, (x_i, y_i) as follows:

$$d_i = \sqrt{(x_i - x_f)^2 + (y_i - y_f)^2} \tag{4.27}$$

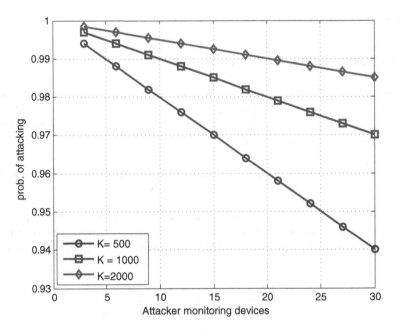

Fig. 4.7 Fake point sub-scheme attacking probability

Since $d_i, i = 1, 2, \ldots, m$ has the same value and (x_f, y_f) is a fixed point for all MNNs, then the attacker calculates the same estimated location (x_e, y_e) for an MNN$_i$'s true location (x_i, y_i). Hence, (4.4) is expressed as follows:

$$Z_i = d_i + \eta_i + \delta_i, \delta \geq 0 \tag{4.28}$$

where δ_i is a deviation of the estimated distance that is added when applying the fake point sub-scheme. It can be calculated as follows:

$$\delta_i = \frac{|y_i - y_e|}{|x_i - x_e|} \tag{4.29}$$

Therefore, (4.4) changes as follows:

$$Z = d + \eta + \delta, \delta \geq 0 \tag{4.30}$$

The attacker then uses ML estimation as in (4.13) to determine the MNN's position as follows:

$$\begin{aligned}
\hat{\theta}_{ML} &= \arg \min_{[x,y]^T} \sum_{i=1}^{N} \frac{(\eta_i + \delta_i)^2}{\sigma_i^2} \\
&= \arg \min_{[x,y]^T} \sum_{i=1}^{N} \frac{\eta_i^2}{\sigma_i^2} + \frac{\delta_i^2 + 2\delta_i \eta_i}{\sigma_i^2}
\end{aligned} \tag{4.31}$$

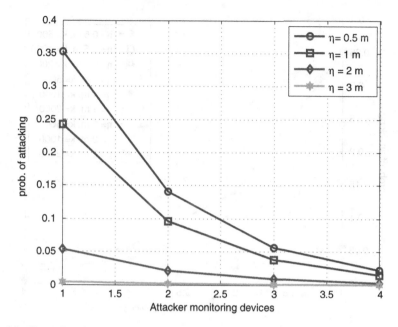

Fig. 4.8 Cluster based sub-scheme attacking probability

where N is the number of an attacker's monitoring devices. Note that the term $\frac{\delta_i^2 + 2\delta_i \eta_i}{\sigma_i^2}$ is the added value to the ML estimation. The additional estimation error is called the correctness of the estimated position and, as it increases, the MNN's location privacy increases as well.

4.6.1.2 Cluster Based Sub-Scheme

The goal of this sub-scheme is to decrease the transmit power by employing transmit power control in such a way that only a small number of an attacker's monitoring devices, L, from all monitor devices, N, can detect the MNN's signal and measure the RSS. Therefore, the attacker calculates the ML estimation as follows:

$$\hat{\theta}_{ML} = \arg \min_{[x,y]^T} \sum_{i=1}^{L} \frac{(z_i - d_i)^2}{\sigma_i^2} \qquad (4.32)$$

For $L = 1$, which means only one monitoring device can detect the MNN's signal, (4.32) can be written as:

$$\hat{\theta}_{ML} = \frac{(\eta)^2}{\sigma^2} + \delta_{cluster} \qquad (4.33)$$

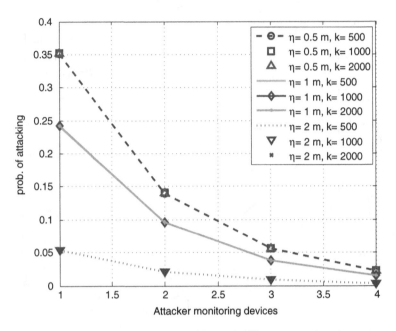

Fig. 4.9 Fake point- Cluster based scheme attacking probability

where $\delta_{cluster}$ is the added estimation error resulting from the lack of information as only one monitoring device measures the MNN's RSS. $\delta_{cluster}$ decreases as the number of monitoring devices increases, which in turn, gives an indication of a lower location privacy level.

4.6.1.3 Fake Point-Cluster Based Scheme

Depending on the analysis of both fake point and cluster based sub-schemes, the estimation error for the combination of the two sub-schemes can be expressed as follows:

$$\hat{\theta}_{ML} = \arg \min_{[x,y]^T} \sum_{i=1}^{L} \frac{(z_i - d_i)^2}{\sigma_i^2} \frac{\delta_{cluster}}{L}$$

$$= \frac{\delta_{cluster}}{L} \arg \min_{[x,y]^T} \sum_{i=1}^{L} \frac{\eta_i^2}{\sigma_i^2} + \frac{\delta_i^2 + 2\delta_i \eta_i}{\sigma_i^2} \qquad (4.34)$$

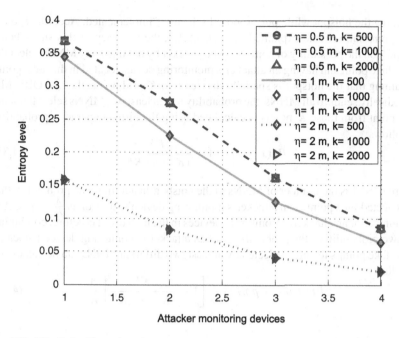

Fig. 4.10 Fake Point Cluster based entropy

4.6.2 Accuracy

In this sub-section, the accuracy of the fake point-cluster based scheme is calculated by measuring the accuracy of the positioning system employed by the attacker. We measure the accuracy of the positioning system by calculating the probability of

Table 4.1 NEMO-based VANET simulation parameters

Road width	$5500\,m \times 10\,m$
Road's network size	$1000\,m \times 10\,m$
Road's networks number	6
Spatial grid points, k	1000
Vehicles number	36000
WiFi size	$45\,m \times 45\,m$
Wi-Fi nodes number	600
Frequency	$2.4\,GHz$
AP transmission power	$5\,mW \simeq 7\,dBm$
cluster area	$25\,m^2$
AP required received power	$5\,dBm$
Cluster required received power	$3\,dBm$
overlapping area among clusters	$1\,m$
Length of the phy header	$0\,byte$
Thermal noise	$0\,dB$

attacking the hotspot while our proposed scheme is implemented. According to the fake point sub-scheme explained in Sect. 4.6.1.3, the accuracy of this sub-scheme depends on the possibility of confusing the attacker by having many MNNs select the same fake point and having an attacker's monitoring device located in this fake point. Assuming that the hotspot is spatially divided into K grid points that the OBU/MR periodically sends to all MNNs, the probability that at least two MNNs select the same fake point from those K points is calculated using the birthday paradox probability as follows:

$$Pr(x \geq 2) = 1 - \frac{K!}{(K-u)!K^u} \tag{4.35}$$

where $u > 1$ is the number of MNNs in the hotspot. In addition, the probability that the selected fake point is an attacker's monitoring device's location is $\frac{A}{K}$, where A is the number of an attacker's monitoring devices in the network. Therefore, combining the two probabilities, the probability that an attacker's monitoring device is located at the fake point's location selected by at least two different MNNs can be calculated as follows:

$$Pr(fake-point) = \left[1 - \frac{K!}{(K-u)!K^u} \right] \frac{A}{K} \tag{4.36}$$

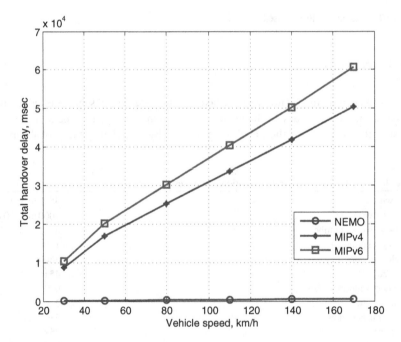

Fig. 4.11 Total handover delay

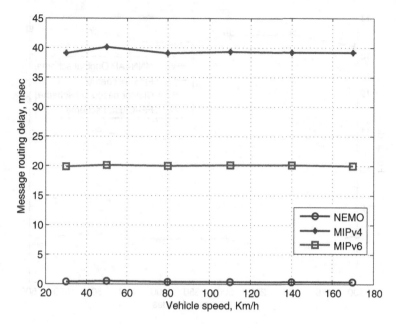

Fig. 4.12 Message routing delay

Since the number of passengers inside the hotspot is always much less than the defined spatial grid points($u \ll K$), we consider $\frac{K!}{(K-u)!K^u} \approx 0$. Therefore, the probability of successfully attacking the hotspot when employing the fake point sub-scheme (Fig. 4.7) is calculated as follows:

$$Pr(fake - pointattacking) = 1 - \frac{A}{K} \qquad (4.37)$$

As illustrated in the figure, the probability of attacking decreases when the ratio (A/K) of the number of the attacker's monitoring devices, A, to the number of defined spatial grid points, K, increases, because the possibility that the selected fake point is an attacker's monitoring device's location increases and hence the attacker is confused. Intuitively, this ratio increases when A increases and/or K decreases.

For the cluster-based sub-scheme, the number of overlapping clusters, O, that intersect with the MNN's cluster affects the probability of achieving an MN's location privacy. This probability is calculated as follows:

$$Pr(cluster) = \Pi_{i=1}^{O} P_{\eta_i}$$

$$= \Pi_{i=1}^{O} \frac{1}{\sqrt{2\pi}\sigma_i} \exp\left(-\frac{\eta_i^2}{2\sigma_i^2}\right) \qquad (4.38)$$

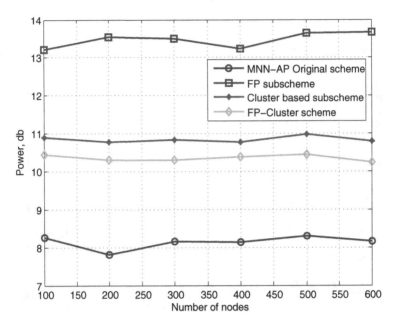

Fig. 4.13 MNN transmission power

As shown in Fig. 4.8, we define the maximum number of overlapping clusters, and hence number of attacker monitoring devices, as four.

Combining the fake point with cluster-based probabilities, we get the probability of achieving location privacy with a fake point-cluster based scheme as follows:

$$\left[\left[1 - \frac{K!}{(K-u)!K^u}\right]\frac{A}{K}\right]\Pi_{i=1}^{O}\frac{1}{\sqrt{2\pi}\sigma_i}\exp\left(-\frac{\eta_i^2}{2\sigma_i^2}\right) \qquad (4.39)$$

Figure 4.9 shows the combination of fake point and cluster-based probabilities.

4.6.3 Certainty

An entropy model measures the uncertainty of an attacker's location privacy scheme, calculated as follows:

$$H(x) = \sum_i Pr(x_i)log\frac{1}{Pr(x_i)} \qquad (4.40)$$

Therefore, the entropy for our proposed schemes is depicted in Fig. 4.10.

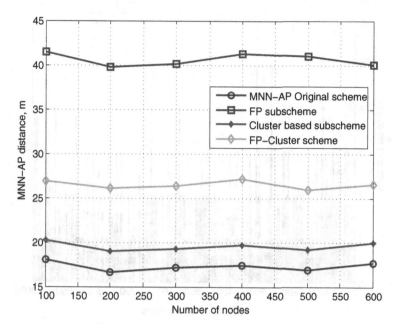

Fig. 4.14 MNN-AP route distances

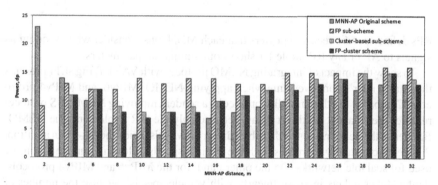

Fig. 4.15 Average power consumption at different distances

4.7 Performance Evaluation

In this section, a simulation has been run to evaluate the performance of the fake point-cluster based scheme. We simulate a 45 m × 45 m hotspot that is installed inside one of the vehicles connected together to create VANET communications. To simulate the overlapping clusters, a group of reference points has been deployed in such a way that each reference point, RP_i, covers an area of 25 m^2, with one meter overlapping area with each neighbor cluster. The centralized AP as well as all

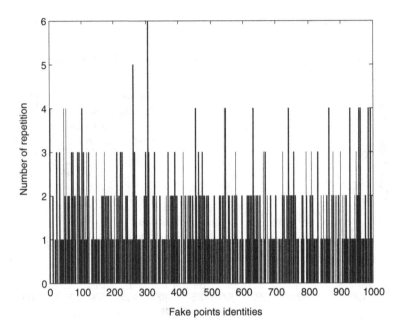

Fig. 4.16 Fake point histogram

RPs define specific received powers that each MNN must consider while sending its signals to AP or any RP. Table 4.1 shows our simulation parameters.

To show the impact of integrating NEMO protocol with VANET, Fig. 4.11 presents the total sender's handover time when applying NEMO, MIPv6, and MIPv4 protocols. Compared to other mobility protocols, a sender employing NEMO BS protocol requires the smallest handover delay, because only the MR implements the NEMO-BS, hence the delaying cost is distributed over all MNNs inside the Wi-Fi. In addition, this delay constantly increases with the vehicle speeds, allowing our scheme to be used for scalable networks. On the other hand, for the MIPv6 and MIPv4 protocols, the handover delays increase linearly with vehicle speeds, because the number of handovers increases accordingly. The MIPv6 protocol costs much more delay than that in the MIPv4 protocol, because of the addition of the correspondent binding update messages transmitted to the sender's correspondent nodes.

Calculating the total message's routing delay, Fig. 4.12 shows that NEMO protocol achieves 95 and 97 % decreases comparing to the delays in MIPv6 and MIPv4, respectively.

Figure 4.13 shows the MNN transmission power for the fake point-cluster based, fake point sub-scheme, cluster based sub-scheme, and the original Wi-Fi communication scheme as a reference. As shown in the figure, the original communication scheme where a fake point-cluster based scheme is not implemented has the smallest transmission power. On the other hand, there is a 65.5 % power increase when employing the fake point sub-scheme because the selected fake point can be found

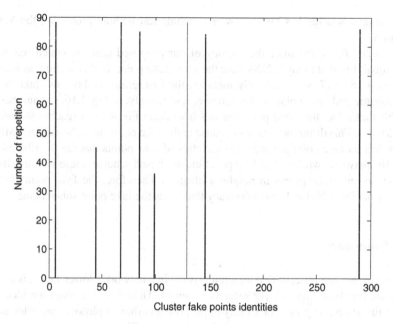

Fig. 4.17 Fake point cluster based histogram

very far from the MNN, thus more power at MNN is needed to equalize RSS at this fake point. The power required in the cluster-based sub-scheme depends on the received power at the RP, which is always less than the received power at the AP; therefore, only a 37.5 % increase in MNN transmission power is recorded. Compared to the fake point sub-scheme, when combining the fake point sub-scheme with the cluster-based sub-scheme, we get a 23 % decrease in the transmission power. The reason for this power saving is that when employing the fake point-cluster based scheme, the MNN selects a nearer fake point, located in its cluster.

The distances between MNNs and APs contribute to increasing MNNs' transmission power, as illustrated in Fig. 4.14. The shortest distance between MNN and AP, which is employed in an MNN-AP conventional scheme, always requires less transmission power, while the indirect distances from an MNN to the fake point then to the AP, which are employed in our proposed schemes, consume more power. Compared to the reference MNN-AP conventional scheme, the fake point, cluster-based, and fake point-cluster based schemes encounter distance increases of 135, 17.6, and 52.9 %, respectively.

The increases in distances and powers are our cost to achieve high MNN location privacy. Figure 4.15 shows power consumed at different MNN-AP distances. Our proposed schemes achieve lower power consumptions than that in the conventional scheme at MNN-AP distances less than 5 m. At such small distances, the location protection is much more important than it is at the large distances where MNNs locations can be revealed easily. Therefore, at lower distances, the fake point-cluster scheme achieves both less power consumption and high location privacy, while the conventional scheme has higher power consumption without protecting the MNN location.

In Sect.4.6, we calculate the entropy of our proposed scheme, which relies on the probability that many MNNs have the same fake point. In this section, as shown in Figs.4.16, 4.17, we practically measure the histogram for both the fake point sub-scheme and fake-point cluster scheme, respectively. In Fig.4.16, the number of MNNs that select the same point reaches 6 while in Fig.4.17, it reaches 90 out of 300 MNNs. This difference occurs because in the fake point sub-scheme, each MNN can select its fake point among large varieties of fake points that are distributed all over the network, while in the fake point-cluster based scheme, these varieties have shrunk to only fake points in neighbor clusters. Therefore, the fake point-cluster scheme achieves higher location privacy than does the fake point sub-scheme.

4.8 Summary

In this chapter, we observe that location privacy in a network's lower-layers is a pre-requisite for those higher-layer security schemes. Therefore, we adapt the ideas of the obfuscation and power variability to propose an efficient physical-layer location privacy scheme, the fake point-cluster based scheme. The proposed scheme thwarts physical-layer attackers occurring inside a public hotspot in a moving vehicle, in order to ensure mobile network nodes' location privacy. Employing the correctness, accuracy, and certainty as metrics, we analytically measure the location privacy achieved by our proposed scheme. In addition, using extensive simulations, we evaluate the performance of the hotspot in NEMO-based VANET when employing our proposed scheme comparing to the traditional hotspots. Furthermore, we show that our proposed scheme can practically be implemented, due to the possibility of having at least two nodes select the same fake point.

References

1. Lorchat, J., Uehara, K.: Optimized inter-vehicle communications using nemo and manet. In: Proceedings of the Third Annual International Conference on Mobile and Ubiquitous Systems: Computing, Networking, and Services, pp. 1–6. IEEE, San Jose, CA, USA (2006)
2. Baldessari, R., Festag, A., Abeillé, J.: Nemo meets vanet: a deployability analysis of network mobility in vehicular communication. In: Proceedings of the 7th International Conference on Intelligent Transport Systems Telecommunications (ITST'07), pp. 1–6. IEEE, Sophia Antipolis, France (2007)
3. Céspedes, S., Shen, X., Lazo, C.: Ip mobility management for vehicular communication networks: challenges and solutions. IEEE Commun. Mag. **49**(5), 187–194 (2011)
4. Prakash, A., Tripathi, S., Verma, R., Tyagi, N., Tripathi, R., Naik, K.: Vehicle assisted cross-layer handover scheme in nemo-based vanets (vanemo). Int. J. Internet Protoc. Technol. **6**(1), 83–95 (2011)
5. Gezici, S.: A survey on wireless position estimation. Wirel. Person. Commun. **44**(3), 263–282 (2008)

6. El-Badry, R., Sultan, A., Youssef, M.: Hyberloc: providing physical layer location privacy in hybrid sensor networks. In: Proceedings of the IEEE International Conference on Communications (ICC 2010), pp. 1–5. IEEE, Cape Town, South, Africa (2010)
7. El-Badry, R., Youssef, M., Sultan, A.: Hidden anchor: a lightweight approach for physical layer location privacy. J. Comput. Syst. Netw. Commun. 2010, 1–12 (2010)
8. Ardagna, C., Cremonini, M., Damiani, E., De Capitani di Vimercati, S., Samarati, P.: Location privacy protection through obfuscation-based techniques. In: Proceedings of the Data and Applications Security XXI, pp. 47–60. Springer, Redondo Beach (2007)
9. Jiang, T., Wang, H.J., Hu, Y.C.: Preserving location privacy in wireless lans. In: Proceedings of the 5th International Conference on Mobile Systems, Applications and Services, MobiSys '07, pp. 246–257. ACM, San Juan, Puerto Rico (2007). doi:10.1145/1247660.1247689. http://doi.acm.org/10.1145/1247660.1247689
10. Wang, T., Yang, Y.: Location privacy protection from rss localization system using antenna pattern synthesis. In: Proceedings of the IEEE INFOCOM 2011, pp. 2408–2416. IEEE, Shanghai, China (2011)
11. Huang, L., Matsuura, K., Yamane, H., Sezaki, K.: Enhancing wireless location privacy using silent period. In: Proceedings of the IEEE Wireless Communications and Networking Conference, WCNC 2005, vol. 2, pp. 1187–1192. IEEE, Orleans, USA (2005)
12. Oh, S., Vu, T., Gruteser, M., Banerjee, S.: Phantom: Physical layer cooperation for location privacy protection. In: Proceedings of the IEEE INFOCOM 2012, pp. 3061–3065. IEEE, Orlando, FL, USA (2012)
13. Taha, S., Shen, S.: A link-layer authentication and key agreement scheme for mobile public hotspots in NEMO based VANET. In: Proceedings of the Communication and Information System Security Symposium (Globecom12 CISS). Anaheim, USA (2012)
14. Sun, Y., Lu, R., Lin, X., Shen, X., Su, J.: An efficient pseudonymous authentication scheme with strong privacy preservation for vehicular communications. IEEE Transact. Veh. Technol. 59(7), 3589–3603 (2010)
15. Lu, R., Lin, X., Zhu, H., Ho, P., Shen, X.: A novel anonymous mutual authentication protocol with provable link-layer location privacy. IEEE Transact. Veh. Technol. 58(3), 1454–1466 (2009)
16. Lin, X., Sun, X., Ho, P., Shen, X.: Gsis: A secure and privacy-preserving protocol for vehicular communications. IEEE Transact. Veh. Technol. 56(6), 3442–3456 (2007)
17. Armknecht, F., Girao, J., Matos, A., Aguiar, R.: Who said that? privacy at link layer. In: Proceedings of the 26th IEEE International Conference on Computer Communications, INFOCOM 2007, pp. 2521–2525. IEEE, Anchorage, AK, USA (2007)
18. Ryu, E., Yoon, E., Yoo, K.: More robust anonymous authentication with link-layer privacy. In: Proceedings of the IEEE Asia-Pacific Services Computing Conference, APSCC 2010, pp. 441–446. IEEE, Hangzhou, China (2010)
19. Fan, Y., Lin, B., Jiang, Y., Shen, X.: An efficient privacy-preserving scheme for wireless link layer security. In: Proceedings of the IEEE Global Telecommunications Conference, GLOBECOM 2008, pp. 1–5. IEEE, New Orleans, USA (2008)
20. Lu, R., Lin, X., Zhu, H., Ho, P., Shen, X.: Ecpp: Efficient conditional privacy preservation protocol for secure vehicular communications. In: Proceedings of the 27th IEEE Conference on Computer Communications, INFOCOM 2008, pp. 1229–1237. IEEE, Phoenix, USA (2008)
21. Katev, P.: Propagation models for wimax at 3.5 ghz. In: Proceedings of the ELEKTRO 2012, pp. 61–65. SLOVAKIA (2012). doi:10.1109/ELEKTRO.2012.6225572
22. Shokri, R., Theodorakopoulos, G., Le Boudec, J., Hubaux, J.: Quantifying location privacy. In: Proceedings of the IEEE Symposium on Security and Privacy, SP 2011, pp. 247–262. IEEE, Oakland, CA, USA (2011)
23. Ezzine, R., Al-Fuqaha, A., Braham, R., Belghith, A.: A new generic model for signal propagation in wi-fi and wimax environments. In: 1st IFIP on Wireless Days, WD'08, pp. 1–5. IEEE, Dubai, United Arab Emirates (2008)

Chapter 5
Conclusions and Future Directions

In this chapter, we present the main research results and discuss future work.

5.1 Conclusions

Having the goal of securing a mobility management protocol that supports mobile nodes with seamless communications, this research has proposed three security and privacy schemes employed in three different mobility management for VANET scenarios. In our proposed schemes, we have considered the VANET mobility management challenges, including high speed vehicles, multi-hop communications, and scalability.

In Chap. 2, based on onion routing and anonymizer, we have proposed an anonymity and location privacy scheme, ALLP, to be employed in MIPv6 heterogeneous mobile networks, such as VANETs. With the goal of securing MIPv6 home binding update and return routability signalling messages, the ALPP scheme involves two complementary sub-schemes, anonymous home binding update (AHBU) and anonymous return routability (ARR). In addition, using the certificate-less public key cryptography, we present an efficient and secure authentication scheme between the mobile node and the foreign gateway. The proposed authentication scheme allows the two parties to authenticate each other without constructing a public key infrastructure in the network. Using the entropy model, we compare the achieved degree of anonymity of our proposed scheme with the mix-based scheme. When increasing the number of senders in our system, we observe that the degree of anonymity of our proposed scheme falls between 60 and 85 %, whereas the mix-based scheme encounters a high delay while increasing the degree of anonymity. Regarding the location privacy, we prove that when implementing the ALPP scheme, no network entity except the mobile node itself can identify this node's location, which is represented by the mobile node's care-of address. Furthermore, we prove that when using our proposed key establishment scheme and under the assumption that the attacker knows the mobile node's private key, the attacker is still unable to create the shared key between the mobile node and the foreign gateway. In addition, we also prove that if either the mobile node identity or its public key is changed by an attacker, our proposed key establishment can detect this forgery. Therefore,based on these proofs, we show that the ALPP scheme thwarts traffic analysis, collusion, replay, and man-in-the middle attacks. Using extensive simulations, we show that our proposed sub-schemes decrease the routing delays by 42 % for the AHBU sub-scheme, 43 % for ARR-fixed correspondent node, and 30 % for ARR-mobile node, compared to the triangle routing scheme. In our simulations, we consider different mobile node speeds, and show that routing delays decrease with the high mobility networks, such as VANETs.

In Chap. 3, another scenario for multi-hop PMIPv6 in VANETS has been address, and a novel efficient mutual authentication scheme, EM^3A has been proposed to be employed between a mobile node and a relay node in the multi-hop scenario. In addition, we employ symmetric polynomials to generate a shared key between those nodes. Compared to previous symmetric polynomial schemes that are used to generate a shared key between two arbitrary nodes, our proposed scheme increases the secrecy level achieved in the key establishment from $t-$secrecy to $t \times 2^n-$secrecy. This increase in secrecy level hardens against colluder attacks, to the extent that they require at least $t \times 2^n + 1$ colluders instead of only $t + 1$ colluders to reveal the shared key and break the secure system. Furthermore, we have mitigate the problem of mobile node's revocation by proposing new security steps that also achieve mobile node backward secrecy. Compared to the traditional revocation that changes the whole system keys, our revocation scheme changes only the keys of the mobile nodes sharing the same first-attached mobile access gateway (MAG). However, the trad-off is to store an MAG's list of identities in each authenticated mobile node. By means of simulations, we have shown that EM^3A results in a low delay and allows for seamless communications, even in highly mobile/heavy traffic demanding scenarios. Moreover, we present a case study of a proposed multi-hop authentication PMIP (MA-PMIP) that is implemented in vehicular networks. EM^3A has been implemented as the authentication step in MA-PMIP and compared to the AMA-PMIP scheme that implements the AMA authentication scheme, our MA-PMIP protocol with EM^3A achieves 99.6 and 96.8 % reductions in authentication delay and communication overhead, respectively.

Finally, in Chap. 4, we direct our research to protect the mobile node's location privacy in the physical layer, which is a prerequisite to ensuring location privacy in upper-layers. Relating to our thesis, the scheme proposed in this chapter can be combined with any schemes proposed in Chaps. 2 and 3. In this chapter, we consider the NEMO-based VANET that supports public hotspots installed inside moving vehicles. We have modified the ideas of the obfuscation and power variability to propose the fake point-cluster based scheme, which thwarts physical-layer attacks that exploit a sender's received signal strength to localize this sender. The fake point-cluster based scheme involves two sub-schemes, the fake point and the cluster-based, that together increase the network performance and the achieved location privacy, rather than implementing each one separately. Using correctness, accuracy, and certainty, we have measured the location privacy achieved in our proposed scheme. In addition, using extensive simulations, we evaluate the performance of the hotspot in NEMO-based VANET when employing our proposed scheme compared to the traditional hotspots. Lastly, we show that our proposed scheme can practically be implemented, due to the possibility of having at least two nodes select the same fake point.

5.2 Future Research Directions

Securing mobility management for vehicular ad hoc networks has many challenges. Therefore, many research directions can be outlined to extend our research.

5.2.1 Securing Nested NEMO for Vehicular Ad Hoc Networks

Nested NEMO is a topology type of the NEMO mobile network in which a NEMO mobile router manages the mobility of another mobile routers located in the same mobile network or in neighbor networks. Due to its inefficiency and highly routing delay, such a topology is considered a problem when it occurs in mobile networks, especially with vehicular networks due to their high mobility. Therefore, the trend is to propose a route optimization for nested NEMO, which mobile routers use to communicate with the correspondent nodes instead of directing the messages to all mobile router's home agents. Despite the great number of studies that have been done, there has been no route optimization standard issued yet.

As a next step of our proposal, we will suggest a route optimization scheme to be employed for nested NEMO-based VANETs. Unlike previous schemes, we will consider both the efficiency and the network security for the proposed scheme.

5.2.2 Securing Mobility Management for Electrical Vehicles

During the 1970s and 1980s, the increase of investment in Canada's electrical grid was very noticeable. However, recently, with the overwhelming growth in electricity demand versus supply, coupled with the aging network facilities, catastrophic black-outs, such as the 2012 India blackouts, happen frequently. Therefore, Smart Grid, a worldwide trend, has been developed to support efficient and reliable electricity trans-formation by enabling time-of-use pricing, where electricity prices changes based on the time of the day. The idea is to create two-way communication between con-sumers and suppliers, to transform both the electricity and the information about the consumed power. The customers' interactions, resulting when the supplier supports them with real-time information about their consumed power, allow balancing of the electricity grid in its peak hours, and hence decrease the blackout probability. Benefits of the smart grid include self-healing from power disturbance events, enabling active participation by consumers in demand response, operating resiliently against physical and cyber attack, accommodating all generation and storage options, increasing investment, and optimizing assets and efficient operation.

As an application of the Smart Grid, electric vehicles increase the benefits, including lower operating and maintaining cost, reduced vehicle noise and making the "green" choice, and increased energy efficiency. Therefore, the Canadian government's vision is to have one electric vehicle among every 20 vehicles by 2020. The battery of the electrical vehicle is powered by plugging it in to the electricity grid in order to store energy. Accordingly, a new communication, namely Vehicle to Grid (V2G), has been considered recently. Furthermore, supporting seamless communication in V2G, where the activated IP connection of the electric vehicle is not interrupted while the vehicle is moving through different networks, is required.

In our future work, the security and privacy of the electrical vehicle in such communication will be considered. The smart grid uses Bad Data Detector (BDD) techniques to indicate erroneous measurements, however false data injection attacks could carefully change a selected set of measurements to add arbitrary errors in the state estimations without triggering the BDD's alarm. In addition, unauthorized customers and fake suppliers could steal electricity and deceive other entities in the grid. Moreover, privacy attackers may analyze the real-time transmitted data about customers and maliciously reveal customers' habits and locations.